PHOTOVOLTAIC SAFETY

AIP CONFERENCE PROCEEDINGS 166
RITA G. LERNER
SERIES EDITOR

PHOTOVOLTAIC SAFETY

DENVER, CO 1988

EDITOR:
WERNER LUFT
SOLAR ENERGY RESEARCH INSTITUTE

AMERICAN INSTITUTE OF PHYSICS NEW YORK 1988

L.C. Catalog Card No. 88-42854
ISBN 0-88318-366-8
DOE CONF-880112

Printed in the United States of America.

CONTENTS

Foreword

These proceedings contain the papers that were presented at the Photovoltaics Safety Conference held on 19–20 January 1988 in Denver, Colorado. The conference was sponsored by the Solar Energy Research Institute of Golden, Colorado.

This was the second photovoltaic safety conference sponsored by the Solar Energy Research Institute. The previous conference was held on 16–17 January 1986 in Lakewood, Colorado. The proceedings from that conference were published in *Solar Cells,* Vol. 19, Nos. 3–45 (January 1987).

Research, development, and production of photovoltaic devices, especially thin-film devices, involves the use of toxic and hazardous chemicals, small amounts of which can pose a serious health hazard and, in some cases, can be lethal. Thus the health and safety of personnel engaged in these activities and of the people in communities surrounding photovoltaic research and production facilities must be protected by careful attention to protective safety measures.

The objectives of this conference were to exchange information on safety measures used by parties engaged in research, development, and production of photovoltaic devices, to describe developments since the 1986 photovoltaic safety conference, to explore substitution of less hazardous materials and process alternatives, to identify health and safety issues needing further analysis, and to discuss avenues to maintain the good safety record enjoyed by the photovoltaic community.

For this purpose, the conference was organized into six sessions that addressed alternative processes, health issues, waste handling, feedstock handling, laws and regulations, and other issues. The conference was attended by 79 persons from private industry, universities and research laboratories, U.S. national laboratories, representatives from the U.S. Department of Energy, and foreign organizations in Germany, Japan, India, Great Britain, and The Netherlands.

This successful conference was possible because of the expert help from Diane Christodaro, SERI's Conference Coordinator, and Patricia Haefele and Kathy Summers, SERI.

Werner Luft
Conference Organizer and Chairman
Solar Energy Research Institute

January 1988

Introduction

Lloyd O. Herwig
Photovoltaic Energy Technology Division
U.S. Department of Energy
Washington, D.C. 20585

There is no more important research and discussion topic than the safety and understanding of photovoltaic materials and operational procedures relative to activities in research laboratories and pilot and production plants, and to protection of the general public safety and environmental well-being, both on-site and off-site. A profound respect for and interest in the health and safety aspects of photovoltaic technology is evident in the attendance and support of so many persons and organizations at this conference and at the previous conference in January 1985. The photovoltaic community continues to give close attention to the environmental, health, and safety (EH & S) aspects of the entire range of activities involved in this emerging commercial technology, from research to production facilities. This commitment arises in part from the fact that photovoltaic energy conversion (and power production) is essentially pollution-free in its basic function, and thus is expected to introduce minimal EH & S impacts in the generation of power. In keeping with this clean image, it behooves the community to take all relevant actions possible to maintain that image, from research, through fabrication and construction, to large-scale commercialization.

Our concern and attention in this area must be continuous and ingenious, since the approach to acceptable personal and public risk is largely an inductive methodology. That is, no matter how thoroughly one may try to think through all of the possibilities for trouble and how many automatic and operational precautions (and interlocks) might be incorporated, it is possible that some unanticipated sequence of events could still lead to potential problems. Even though the chemical and electronics industries have faced and solved many similar types of problems, their approaches must be a starting point in undertaking independent assessments of the needs of photovoltaic technology. Consideration of EH & S aspects can never been taken for granted (or complete).

Thus environmental, health, and safety considerations must receive the best efforts of all participants in designing, siting, and constructing facilities; selecting source materials, fabrication methods, and operating procedures; evaluating potentially hazardous conditions; educating the participants and outside community services and leaders; and communicating openly within the photovoltaic community about EH & S approaches and experiences.

The U.S. Department of Energy is eager to facilitate progress in this area, and is open to suggestions concerning its ongoing role; for example, its support of this workshop series (as planned and carried out by the Solar Energy Research Institute); its encouragement of SERI and Sandia to provide all practical precautions in their in-house research and in their contractors' research; its support of the Environment, Health, and Safety Assessment Program at Brookhaven National Laboratory to augment the photovoltaic community's efforts; and in its initiative to begin an international dialogue in this area.

IMPLICATIONS OF THE TOXICITY OF TETRAMETHYLTIN, DIMETHYL TIN DICHLORIDE, AND TIN TETRACHLORIDE IN SELECTING A SUITABLE TIN PRECURSOR IN THE CHEMICAL VAPOR DEPOSITION OF TIN OXIDE.

Roy G. Gordon
Harvard University, Cambridge, Mass., 02138

James W. Proscia
BTU Engineering, Corp., N. Billerica, Mass., 01862

Safety in the chemical vapor deposition (CVD) of tin oxide films from tetramethyltin, dimethyltin dichloride, and tin tetrachloride has recently become a matter of much concern. In deciding upon a possible tin precursor, the potential health hazards have to be balanced against the benefits to solar cell fabrication. If one of the more toxic precursors is found desirable, appropriate precautions need to be taken to minimize exposure.

Tetramethyltin is the most toxic of the tin precursors. This compound is found to cause irreversible central nervous damage probably after conversion to trimethyltin in the liver. Tin tetrachloride is not believed to have any toxicity related to the tin, but is irritating due to its acidic character. Tetramethyltin and tin tetrachloride are both liquids with fairly high vapor pressures. Spills of these two chemicals are therefore somewhat more difficult to deal with than dimethyltin dichloride which is a solid.

The proper treatment of the exhaust by-products in the CVD deposition of tin oxide from each of these tin precursors is important. The exhaust of the tetramethyltin process can be burnt in an after-burner. The dimethyltin dichloride may be treated by an after-burner and then a water scrubber because of the generation of hydrochloric acid. The tin tetrachloride must be treated by a scrubber.

In solar cell fabrication, all three tin precursors will produce suitable tin oxide with respect to their transmission, sheet resistance, and roughness. With respect to cost, tetramethyltin is much more expensive than dimethyltin dichloride and tin tetrachloride.

INTRODUCTION

A properly designed transparent conducting electrode is an important component of an amorphous silicon solar cell. Such an electrode should have as high a conductivity as possible while not absorbing very much visible light. Typically sheet resistances of between 8 to 10 $\Omega/\square$ with about 10% visible absorption are satisfactory for this application. Of additional concern is the incorporation of sufficient texturing into the electrode to enhance the light trapping effect. Fluorine doped tin oxide has been widely accepted as the material of choice for this contact. Tin oxide is deposited by atmospheric chemical vapor deposition (CVD) from tin tetrachloride - $SnCl_4$, tetramethyltin - $(CH_3)_4Sn$, and dimethyltin dichloride - $(CH_3)_2SnCl_2$.

Recently there has been much concern about the safety of the chemical precursors for the deposition of tin oxide films. Tetramethyltin is considered highly toxic [1], dimethyltin dichloride is moderately toxic [2], and tin tetrachloride [3] is not believed to be toxic but is corrosive. In choosing a suitable precursor, the various health risks have to be weighted against such aspects of the deposition as the quality of the films produced, the cost of the process for each precursor, and special equipment demands. Once a choice is made, the safety and health risks of each process will have to be addressed.

TOXICITY

The toxicity of organotin compounds is generally agreed to follow the following order:

$$SnCl_4 < R_3SnCl < R_2SnCl_2 < R_4Sn < R_3SnCl$$

where R is an alkyl group. In general, the smaller substituents such as a methyl group result in higher toxicity than larger groups such as butyl. Tin tetrachloride is reported to have no toxicity related to the tin, but its acidic nature may damage the eyes and mucous membranes. Tetramethyltin is quite toxic with a (intraperitoneal) LD_{50} of 18 mg/kg in rats [1]. Tetramethyltin is found to cause irreversible central nervous system (CNS) damage probably after conversion to trimethyltin in the liver. Exposure to trimethyltin has been documented to cause long-term CNS damage in humans [4]. The oral LD_{50} of the dimethyltin dichloride is higher (74 mg/kg) and it is therefore safer than tetramethyltin.

Though dimethyltin dichloride is less toxic than tetramethyltin, the OSHA required TLV values for both compounds are 0.1 mg $(Sn)/m^3$. The TLV for tin tetrachloride is over an order of magnitude higher with a value of 2.0 mg $(Sn)/m^3$ [5].

THE CHEMICAL REACTIONS AND PHYSICAL PROPERTIES

The cost of the tin oxide precursors is an important consideration is selecting a process. Currently tetramethyltin costs $43./(mole tin) in bulk, dimethyltin dichloride $2.90/(mole tin), and tin tetrachloride about $1.70/(mole tin). In the large scale production of tin oxide coated glass, a conveyorized belt furnace with a suitably designed reaction zone is used (figure 1). In this reaction zone reacting gases are introduced, allowed to react, and then exhausted. Glass substrates continuously move through the reaction zone and are subsequently coated with the tin oxide. Of considerable importance is the injector head used to introduce the gases. This injector must introduce the gases uniformly across the glass. The design of this injector is dependent on the specifics of the reaction being used. It is recommended that the reaction region and the entrance and exit of the belt furnace be placed within a high quality hood. The hood over the reaction region will provide a degree of safety should the injector head develop a leak. In normal operation gases from the reaction region do not reach either the entrance or exit of the furnace. However, tetramethyltin and dimethyltin dichloride are both flammable and a sudden ignition and explosion of the gases could cause gases to stream out the end of the furnace.

The CVD deposition of tin oxide from tetramethyltin and dimethyltin dichloride is a controlled oxidation described by the following equations:

$$(CH_3)_4Sn + O_2 + \text{dopant} \rightarrow SnO_{2-x}F_x + CO_2 + H_2O + \text{other by-products} \tag{1}$$

$$(CH_3)_2SnCl_2 + O_2 + \text{dopant} \rightarrow SnO_{2-x}F_x + CO_2 + H_2O + HCl + \text{other by-products} \tag{2}$$

In both of these reactions, the preferred dopant is bromotrifluoromethane. Since both of these tin precursors do not react with oxygen at room temperature, the reacting gases may be mixed prior to deposition. This allows considerable simplification of the injector used in the depositions. The injector head need only use a single set of holes.

The deposition of tin oxide from tin tetrachloride proceeds via the following equation:

$$SnCl_4 + \text{hydroxylated compound} + \text{dopant} \rightarrow SnO_{2-x}F_x + HCl + \text{other by-products} \tag{3}$$

The hydroxylated compound is typically water or an alcohol. In this reaction, the tin tetrachloride reacts with either the water or alcohol at room temperature, thereby requiring that the reactants be mixed together just prior to deposition. This requirement makes the design of an injector head somewhat more complicated.

The three tin precursors may be introduced to the injector heads as vapors by passing dry nitrogen gas through a bubbler. Tetramethyltin and tin tetrachloride are liquids at room temperature with sufficient vapor pressure to achieve the necessary concentrations. Dimethyltin dichloride is a solid at room temperature. This compound must be heated to approximately 130 ^{o}C at which temperature it is a liquid and has sufficient vapor pressure. The bubbler system is somewhat more cumbersome for dimethyltin dichloride because it is a solid at room temperature. This compound must first be melted in order to recharge the bubblers.

At room temperature the vapor pressures of tetramethyltin, tin tetrachloride, and dimethyltin dichloride are 100, 30, and << 0.1 mm Hg, respectively. Because tetramethyltin is a high vapor pressure liquid, the toxic effects of this compound are somewhat magnified in the event of a spill. This compound has only a slight hydrocarbon-like odor which does not give adequate warning of exposure. Spills of this compound would be extremely difficult to contain and therefore evacuation of the vicinity of the spill is necessary. The spilled liquid should be allowed to dissipate and adequate ventilation employed. Tin tetrachloride is also a high vapor pressure liquid, but its vapors have an irritating burning odor. Spills of this material can be controlled by pouring water or an alkaline solution on the affected region. Personnel attempting to treat large spills of either tin tetrachloride or tetramethyltin should use a self-contained breathing apparatus. Spills of dimethyltin dichloride are perhaps the easiest to deal with because this compound is a solid which can be quickly swept up. The odor of dimethyltin dichloride is quite putrid and offers some degree of warning to a possible spill (the odor however is not an indication of exposures as low as the TLV)

Of considerable concern for all of these processes is the proper treatment of the reaction by-products. The by-products and residual reactants from tetramethyltin and dimethyltin dichloride should be passed through an after-burner in which any residual reactants are burned. Unreacted tin compounds are converted to tin oxide in the after-burner. The exhaust from the dimethyltin dichloride should be further treated with a water scrubber containing a basic solution in which the hydrochloric acid generated in this process is neutralized. Tin tetrachloride cannot be burnt, and therefore the exhaust from this process must be neutralized in a water scrubber. Of additional concern is the proper disposal of the water effluent flowing from the scrubber. Typical allowable tin cation concentrations in water draining into a sewer system are about 1 ppm [6]. Therefore, the scrubber should be maintained at a pH of 8 to 9 so that the tin is precipitated as the hydroxide, which can be collected and disposed of in a landfill [6,7]. In designing an exhaust system, the corrosive nature of the tin tetrachloride must be taken into account, and the exhaust system line should be constructed from a corrosion resistant material such as Inconel.

THE QUALITY OF THE TIN OXIDE FILMS PRODUCED

All three tin oxide precursors are capable of producing films with conductivities over 2000 $\Omega^{-1}cm^{-1}$. For films grown on soda lime glass, tetramethyltin produces tin oxide films in which the electrical conductivity is not affected very much by the sodium in the glass. However, unless a precoat of silica is used, the nucleation of the films is not complete and large areas of glass are not coated [8]. Dimethyltin dichloride does nucleate uniformly even without a silica precoat, but with some degradation of the conductivity. This sodium effect was found to decrease with increasing film thickness with conductivities of over 2000 $\Omega^{-1}cm^{-1}$ being achieved by a thickness of 6000 Å. For tin tetrachloride, the deleterious effect of the sodium was much more pronounced with films up to a micron showing low conductivities. Therefore, it appears that for

dimethyltin dichloride it might not be necessary to coat the glass with silica. This would lead to a reduction of one step in solar cell fabrication.

Tetramethyltin and dimethyltin dichloride produce textured films at high oxygen concentrations. Films grown at lower oxygen concentrations and higher tin concentrations tend to be smoother. Tin tetrachloride produces textured films over a wide range of conditions with the amount of texturing increasing at higher water vapor concentrations. The amount of texturing necessary for producing a good solar cell is established at a thickness of about 6000 Å while films grown from tetramethyltin and dimethyltin dichloride require a film thickness of about 8000 Å.

Conclusions

It appears that the high cost and high toxicity of tetramethyltin makes this compound unsuitable for the large scale production of tin oxide coated glass. Because of the lower toxicity and cost of dimethyltin dichloride, this compound is more suitable for producing smooth tin oxide films on glass. For textured films, tin tetrachloride or dimethyltin dichloride can be used. With dimethyltin dichloride it should be possible to eliminate coating the glass with a pacifying film such as silica.

1. Chemical Abstracts 46, 6265; W. Zeman, E. Gadermann, and K. Hardebeck, Deut Arch klin Med 198, 713 (1951)

2. J.M. Barness and P.N. Magee, J Pathol Bacteriol 75, 267 (1958)

3. Data sheet for tin tetrachloride, M&T Chemicals Inc, Sheet Number 324

4. W.D. Ross, E.A. Emmett, J. Steiner, and R. Tureen, Am J Psychiatry 138, 1092 (1981)

5. Registry of Toxic effects of Chemical Substances, R.J. Lewis and R.L. Tatken, U.S. Department of Health and Human Services, Washington, D.C., 1979

6. Industrial Pollution Control Handbook, H.F. Lund, McGraw-Hill Book Company, New York, 1971

7. Prudent Practices for Disposal of Chemicals from Laboratories, National Academy Press, Washington, D.C., 1983

8. F.B. Ellis, Jr, A.E. Delahoy, M. Bohm, J.A. Cambridge, and T. Horning, Mat. Res Soc Symp Proc, 70, 557 (1986)

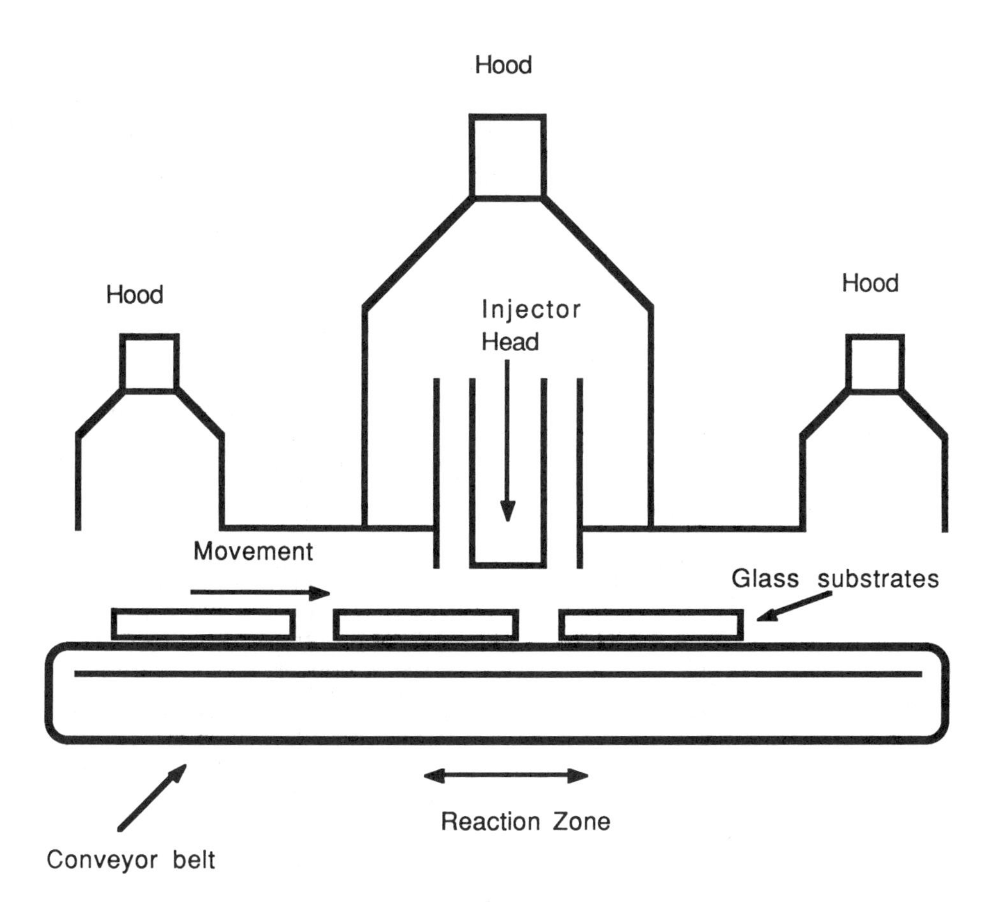
Hood
Hood
Hood
Injector
Head
Movement
Glass substrates
Conveyor belt
Reaction Zone

Figure 1

SAFETY ADVANTAGES OF A CDTE BASED PV MODULE PLANT

M. Doty and P. Meyers
Ametek Applied Materials Lab
352 Godshall Drive
Harleysville, PA 19438

INTRODUCTION

We describe here a process for the manufacture of solar modules based on II-VI compounds which has several advantages over procedures typical of other thin film solar technologies - specifically amorphous silicon. These advantages arise from the fact that all raw materials and waste products are in solid or liquid form. Furthermore, none of the principal raw materials or by-products is flammable, explosive, or acutely toxic [1]. (Nickel, CdO and tellurium are listed by the EPA as acutely toxic, but only in the case of inhalation of fine powders of the first two compounds and of ingestion of the third.) Thus, in both the areas of worker safety and environmental control of hazardous waste, we need be concerned primarily with chronic toxic effects. Discharge of all by-products can be controlled using conventional technology and most metallic wastes can be reclaimed using electrolytic techniques.

In this paper we concentrate on the systems and equipment necessary to eliminate uncontrolled discharge of hazardous materials produced during the fabrication of CdS/CdTe/ZnTe solar modules. It is expected that any operating facility would establish a program for monitoring the level of toxic material in the work environment and in discharged wastes. In addition, it is necessary to establish a program to monitor the long-term exposure of employees and to ensure that proper safety procedures are being followed. These issues, which have been discussed elsewhere [2], are not treated in this work. Furthermore, it is expected that the application and clean-up of various resists and encapsulants will involve organic developers and solvents, the control and reclamation of which are standard practice in the semiconductor industry and will not be covered in detail here.

MANUFACTURING PROCESS

A block diagram outlining the proposed solar cell production process is displayed in Figure 1. The cleaned glass superstrate is first coated with a 0.02 um SiO_2 diffusion barrier, then with a 0.8 um thick SnO_2:F transparent conductive oxide, and then with CdS by pyrolysis in a controlled atmosphere conveyer oven. The n-type CdS is the first of the three semiconductor layers in the nip solar cell structure. CdTe is deposited by electrolysis from an aqueous solution. The composite film is heat-treated in air at 400 C for an hour, then given a mild etch in dilute Br:methanol. The final semiconductor layer, ZnTe:Cu, is deposited by vacuum evaporation onto the heated CdTe surface. Evaporated Ni forms the low resistance contact to the p-type ZnTe. This completes production of the solar cell - in the form of a single large sheet.

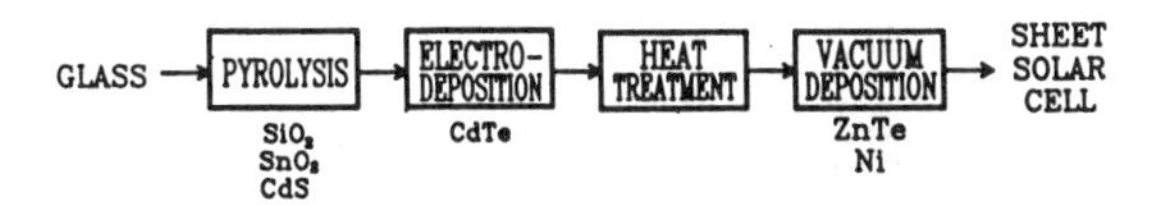

Fig. 1. Block diagram showing the major process steps in the manufacture of CdS/CdTe/ZnTe solar cells.

The process by which a solar cell sheet becomes a module is shown in Figure 2. The single large cell is first patterned into individual solar cells using dry film photolithographic techniques. This is followed by laser scribing of the SnO_2 which electrically isolates the cells. In order to insulate the exposed edges of the solar cell, the entire surface is first covered with an insulating layer of dip-coated SiO_2 which is then patterned in a second photolithographic step. The conductive interconnects, made with sputtered Al, are patterned by laser scribing. After the positive and negative electrical wire contacts are soldered in place, the entire device is spray-coated with a silicone encapsulant.

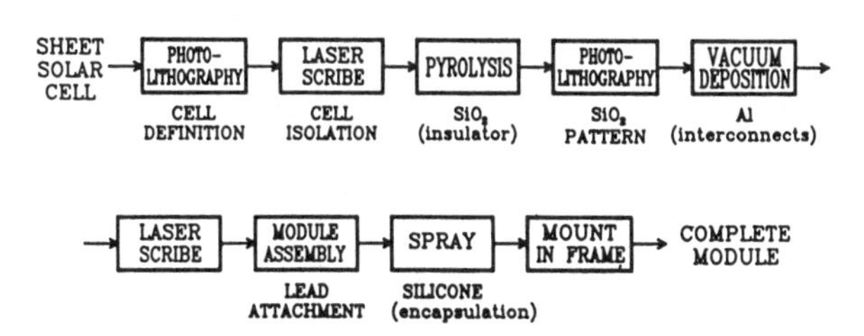

Fig. 2. Block diagram showing the major process steps in the manufacture of CdS/CdTe/ZnTe solar modules.

Table 1. Major Raw Materials

FUNCTION	*COMPONENT*	*RAW MATERIAL*
Superstrate	Glass	Glass
Diffusion Barrier	SiO_2	$Si(OC_2H_6)_4$
TCO	SnO_2:F	$SnCl_2$; $NH_4F \cdot HF$
n	CdS	$CdCl_2$; Thiourea
i	CdTe	Cd; Te
p	ZnTe:Cu	ZnTe powder; Cu
Contact	Ni	Ni
Insulator	SiO_2	$Si(OC_2H_6)_4$
Interconnect	Al	Al
Encapsulant	Silicone	Silicone
Frame, Wires	–	–

PROCESS EFFLUENT AND EMISSION CONTROLS

SnO_2 Deposition

Since the starting material will be low-cost, tempered glass, it will be necessary to protect the SnO_2 from alkali ions which may diffuse from the glass. This will be accomplished by application of a thin SiO_2 diffusion barrier.

Figure 3 is a process flow diagram of these steps. The square-foot glass sheets are cleaned in detergent and rinsed in dilute HCl prior to deposition of the films. We intend to deposit the SiO_2 by pyrolysis of a

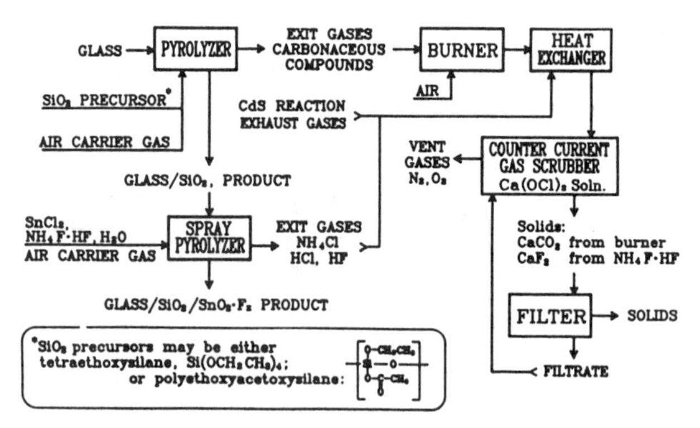

Fig. 3. Process flow diagram showing the deposition of Si_2 and SnO_2 and treatment of by-products.

spun-on or dipped coating of polyethoxyacetoxysilane, tetraethylorthosilicate, or similar compound. This process liberates ethyl alcohol and vapors from incompletely oxidized carbon compounds which should be burned by passing through a flame. The SnO_2:F may be deposited by pyrolysis in air of $SnCl_2$ mixed with NH_4F-HF in an N_2 carrier gas. The by-products from this reaction include fluorides which can be converted to CaF_2 by passing the exhaust gas through a counter-current gas scrubber containing calcium hypochlorite.

CdS Deposition

CdS is deposited by pyrolysis from an aqueous mist comprised of dissolved $CdCl_2$ and thiourea in a nitrogen carrier gas onto a substrate heated to 400 C. The process flow is shown in Figure 4. After passing through the reaction zone, the hot exit gas contains N_2, H_2O vapor, H_2S, HCl, thiourea and its thermal decomposition by-products, $CdCl_2$ and CdO. This exhaust will be cooled with a heat exchanger and then pass into a counter-current gas scrubber. The scrubber solution will contain calcium hypochlorite which will convert the waste aerosol to a suspension of S and CdO particles. The OCl^- will oxidize any ions to harmless species; e.g. $CN^- + OCl^- \rightarrow CNO^- + Cl^-$. The particulates will be filtered out of the scrubber solution while the filtrate will be recycled. Periodically, the collected particulates will be leached in acid to remove the Cd^{++} ions; this solute will then be electrolytically treated as described below.

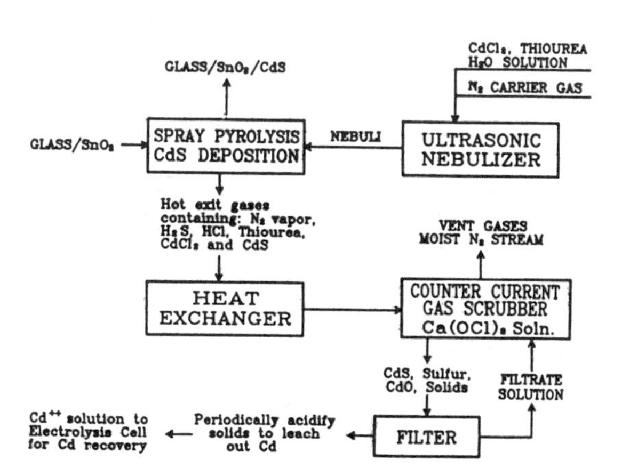

Fig. 4. Process flow diagram showing the CdS deposition system and removal of by-products from the system exhaust.

CdTe Deposition

CdTe is plated at 80 C from an aqueous solution containing approximately one molar Cd^{++} ions and 100 ppm $HTeO_2^+$ ions. Exhaust hoods will be placed over the baths to ensure that any aerosol is drawn out of the work area and can be collected on a filter.

As shown in Figure 5, after plating, the CdTe is rinsed with DI water above the plating bath as it is removed, thus the concentrated rinse water is returned directly to the electrolyte. Then the sample will be given a final DI water rinse and blown dry. The rinse water will be collected in a holding tank. Periodically, the rinse water will be concentrated by boiling, thus providing one source of distilled water for the DI system. The concentrate will be mixed with other solutions containing Cd^{++} and Te^{4+} ions - acid dissolved CdTe from rejected films, spilled electrolyte, etc. - and passed through an exchange column containing Fe, Zn or Cd particles.

We have confirmed in our lab that the Te^{4+} ions (the dominant species in the plating bath) will readily displace the Fe in the acidic medium even at room temperature. There will be some Te^{6+} ions, however, which are created by the oxidizing characteristic of the acidic stripping of CdTe. Removal of these requires that more active metals, higher temperatures, or lower pH be employed. For example, Te^{6+} is readily removed by exchange with Cd metal at pH 2 at 80 C. In any event, the discharged solution will contain primarily Cd ions. This solution will go to an electrolysis cell where the metallic ions can be plated out as metal on the cathode. The plated Cd can be reused in the ion exchange reaction described above. The remaining Cd metal and the elemental Te from the exchange column will be shipped out for reprocessing, and the now dilute solution will be mixed with the final rinse water in the holding tank.

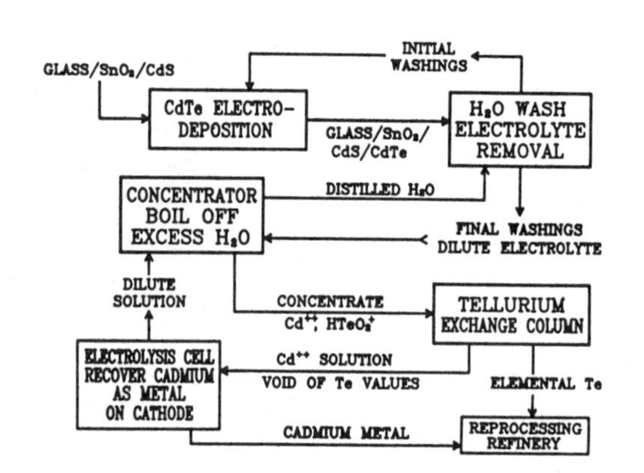

Fig. 5. Process flow diagram of the conversion of dissolved Te^{4+} and Cd^{++} ions into metallic solids.

ZnTe, Ni, and Al Depostion

ZnTe and Cu will be co-evaporated in a vacuum onto a heated substrate in a barrel coater. Evaporated material which is not deposited onto the substrate will build up on the walls of the vacuum chamber. It will therefore be necessary to periodically clean the chamber. This will be done by wiping the interior walls with a cloth soaked in alcohol. The waste-cleaning material including the ZnTe:Cu is inert and can be disposed of with other solid wastes.

Similarly, Ni and Al will be vacuum deposited from solid source material. The only expected by-products are films and flakes of the materials which build up on the inner surfaces of the vacuum chamber. These can be removed with alcohol or dilute acid; disposal is by conventional means.

Photolithographic Processing

Photolithographic processing is required in order to define each solar cell within the module and the insulator pattern. In each case we will apply a dry film resist onto the surface to be patterned. (In the case of the SiO_2 patterning, the film will be applied on the glass side as well.) The standard developing solution is 1:1:1 trichloroethane; the standard stripping solution is methylene chloride. The developing and stripping operations will be performed in a fume hood which would include a cooled baffle to condense most of the vapors and a carbon filter to absorb the rest. The spent solutions can then be purified for re-use by distillation.

The cell patterning, in which cells are defined by removal of Ni, ZnTe, CdTe, and CdS layers in selected areas, is performed with a stripping solution composed of a mixture of HCl and H_2O_2. An alternative is to strip the films in HNO_3. The spent stripping solution will then be processed for Cd and Te recovery as described above.

The individual cells will be electrically isolated by laser scribing through the SnO_2. This process is expected to generate some fine particulates which will be blown away from the surface with N_2 . The scribing chamber will have an exhaust duct so that the particles, which will be SnO_2 mixed with a small amount of glass, can be mechanically filtered out of the exhaust stream.

As was described above, the SiO_2 layer will be deposited by the dip and fire process and the pattern will be defined photolithographically. In this case the stripping solution must remove only SiO_2 and <u>not</u> attack Ni, SnO_2, ZnTe, CdTe, or CdS. This is accomplished with a

Table 2. Environmental considerations

PROCESS STEP	*WASTE PRODUCT*	*TREATMENT*
Deposit SiO_2	Ethanol, Unburned Carbon Compounds	Burn
Deposit SnO_2	SnO_2	Counter Current Gas Scrubber - Neutralize w/Calcium Hypochlorite
Deposit CdS	CdS; HCl; H_2S; $CdCl_2$; Thiourea	same
Deposit CdTe	Scrap CdTe; Waste Electrolyte; Rinse water	Electroless & Elecro-lytic recovery of Cd and Te from conc. solution
Deposit ZnTe; Al; Ni	ZnTe; Al; Ni	
Clean, Rinse	Ethanol, Methanol, Detergents	Reclaim organic solvents
Photo Resist	1:1:1 trichloroethane, methylene chloride	Reclaim organic solvents

buffered HF solution. The spent solution will be treated by mixing with CaO which converts the F^- to CaF_2.

Encapsulation

The module will be coated on the back with a silicone encapsulant. Formulations made specifically for photovoltaic modules are available. The silicone as supplied is a liquid which can be sprayed onto the completed module inside a fume hood. The resin may be cured in air at room temperature, but often heat or chemical accelerators are used to shorten the cure time. During spraying and curing, fumes from the diluents are emitted. These consist of 1:1:1 trichloroethane and methylene chloride. Neither material is flammable. These may be reclaimed by collection on a cooled baffle in the exhaust followed by absorption in carbon filters.

CONTINGENCIES FOR CATASTROPHIES

In general, the personnel and the environment are in relatively little danger in the event of a catastrophe because

1. All materials are in either liquid or solid form and are therefore not easily transported from the location of the manufacturing facility.

2. All materials are used in relatively small quantities and will therefore be quickly dispersed or diluted as they are carried from the scene.

3. All materials are harmful only in the event of ingestion, inhalation or injection into the body and upon long-term exposure (which can be minimized by proper action following the catastrophic occurrence).

Since none of the materials is explosive or pyrophoric, potential hazards occur only in events in which materials are ejected from their containers. This might happen through a leak or spill, explosion, fire or flood. Of these, the explosion and fire have the potential to produce airborne particles which could be inhaled. This would, of course, be a concern in the immediate vicinity of the accident, but would not be expected to significantly elevate the hazard level inherent in such events.

CONCLUSION

Process research on the manufacture of thin film CdTe-based PV modules is an area of active investigation at Ametek AML. While it is expected that there will be refinements and improvements to the procedures described above, we do not foresee modifications which will result in the use or generation of toxic gases or volatile compounds. Since solids and liquids are inherently easier to confine (especially in the event of an industrial accident or natural disaster), it is clear that the safety and environmental protection can be readily incorporated into the final procedures. On the other hand, it is recognized that many of the materials encountered in these processes have the potential to cause serious injury or death. For example, the EPA classifies CdS and CdTe as highly toxic and thiourea as an animal carcinogen. Thus it is necessary to develop procedures which ensure proper control of these and all chemicals and to limit the risk of overexposure - that is the purpose of the procedures outlined above.

Thin film solar cells have the potential to supply power to areas not served by the utility grid and may be especially suitable in developing countries. One method of realizing this potential is to set up manufacturing facilities close to the area in which the modules will be used. These could be locally operated and controlled if the manufacturing technology were within the capability of the local population. We believe that this process not only meets this criterion, but that the production of thin film solar modules based on CdTe can be performed with a high degree of confidence that neither the workers nor the environment will be harmed.

ACKNOWLEDGEMENTS

This work is partially funded through SERI contract ZL-7-06031-2. The authors are grateful to Paul Moskowitz, Brookhaven National Laboratory, for his thoughtful review of the manuscript.

REFERENCES

1. P. D. Moskowitz, P. D. Kalb, J. C. Lee, and V. M. Fthenakis, "An Environmental Source Book on the Photovoltaics Industry", BNL 52052, UC-11, September 1986.

2. S. S. Hegedus, J. D. Meakin, B. N. Baron and J. A. Miller, Solar Cells 19, Vol. 3-4 (1986) pp. 225-236.

LIQUID SOURCES OF ARSENIC FOR THIN-FILM SEMICONDUCTOR DEPOSITION

R.G. Wolfson, Spire Corporation, Bedford, MA 01730

ABSTRACT

There are two chief issues in the evaluation of liquid sources of arsenic for thin-film semiconductor deposition: safety and material quality. Safety concerns center upon toxicity by inhalation, and they favor the use of fully substituted arsenic alkyls; e.g., trimethylarsine and triethylarsine. Considerations of material quality, which reduce finally to the minimization of carbon incorporation, favor mono- and disubstituted alkyl derivatives of arsine; those currently being evaluated are tertiarybutylarsine, diethylarsine, and dimethylarsine.

INTRODUCTION

As conversion efficiencies in thin-film GaAs solar cells have been pushed to record-high values, the toxicity of the arsine used in metalorganic chemical vapor deposition (MOCVD), which is the method of choice for their fabrication, has been recognized as a serious impediment to commercialization [1]. Arsine poses well-documented hazards; it is a highly poisonous gas supplied in high-pressure cylinders, and public concern is properly focussed upon the potential for the widespread dispersal of toxic airborne substances. Despite the remarkably good safety record of the GaAs-based device industry in minimizing the hazards, arsine is becoming the object of legislative prohibition.

The preferred alternative to arsine is a source which is liquid under normal conditions of storage and use. (The identification of a solid source is improbable, since high-vapor-pressure solids are rare; besides, the most likely vapor-phase species would be elemental As_x, which is known to yield excessive carbon incorporation[2]). It can be assumed that any suitable candidate will be made available in bubblers, which obviates most concerns about the reactivity of the source and its potential as a contact poison. This leaves toxicity by inhalation. The precursor reagent itself, in vapor form, must be significantly less toxic than arsine; ideally, the reaction byproducts should be of low toxicity, as well, and should be readily scrubbed from the effluent gas. Further, the gases and particulates which would be produced by the accidental release of either liquid or vapor also should be of low toxicity.

Safety, however, is not the sole consideration, and a replacement for arsine must also satisfy the basic requirements of the technology. The liquid must have a reasonably long shelf life, it must be amenable to purification, and it must be sufficiently volatile for efficient vapor-phase transport. Its stability must be in the proper range: low enough for high-yield decomposition in the reaction zone but high enough to minimize premature cracking. Finally, it must be compatible with the deposition process. This is of extreme importance, since the choice of precursor can alter the growth mechanism and affect the quality of the material deposited.

EXPERIENCE WITH LIQUID SOURCES OF ARSENIC

At the present time, five candidate liquid sources of arsenic are being evaluated:

- trimethylarsine (TMAs) - $(CH_3)_3As$
- triethylarsine (TEAs) - $(C_2H_5)_3As$
- tertiarybutylarsine (TBAs) - $[(CH_3)_3\ C]AsH_2$
- diethylarsine (DEAs) - $(C_2H_5)_2AsH$
- dimethylarsine (DMAs) - $(CH_3)_2AsH$

All of these compounds are available from one or more suppliers, suitably packaged in bubblers, with purities that can best be described as research-grade; <u>i.e.</u>, pure enough for evaluation but not optimized for electronic applications.

The issue of toxicity appears to have been approached empirically, using the established fact that organic derivatives of arsine are less toxic than the molecular hydride itself. The appropriate rule of thumb here is that toxicity becomes much less as alkyl substitution increases. In the absence of other considerations, this would favor TMAs and TEAs, and much of the sparse literature on arsine alternatives for MOCVD deals with these two fully substituted compounds. TMAs is also the only candidate for which toxicological data have been published.

On the other hand, experience with arsine strongly suggests that the hydrogen atoms of the hydride play an important, and perhaps critical, role in MOCVD with TMGa. It is known that arsine and TMGa each accelerates the decomposition of the other, producing methane, and it is surmised that the atomic hydrogen may also scavange reactive methyl radicals adsorbed on the growth surface. This consideration would favor TBAs, DEAs, and DMAs.

The properties of the five alternative As sources are summarized in Table 1. The literature on their use with TMGa for MOCVD is reviewed below.

TABLE 1. Properties of liquid sources of As for GaAs MOCVD

Formula	TMAs[a]	TEAs[b]	TBAs[c]	DEAs[a]	DMAs[d]
Formula Weight	120	162	134	134	106
Melting Temperature	-87°C	-91°C	--	--	-136°C
Boiling Temperature	50-52°C	140°C	65-66°C	101-106°C	37°C
Vapor Pressure	155 torr[e] @ 10°C		96 torr @10°C		
	219 torr[e] @ 18°C			0.801 torr @ 18°C	
	468 torr[e] @ 37°C	15.5 torr @ 37°C			
Toxicity	LC_{50} = 20-22,000 ppm (4 h, mice)				

a Alfa Products, Product Review (1987).

b Texas Alkyls, Product Data (1987).

c Cyanamid, Arsine and Phosphine Replacements for Semiconductor Processing (1987).

d Advanced Technology Materials, Product Data (1987).

e Calculated from the vapor pressure equation.

Trimethylarsine

TMAs (Figure 1) was employed early on for GaInAs MOCVD, where its use avoided the gas-depletion effects caused by the interaction between arsine and the indium alkyl [3]. It pyrolyzes more slowly than arsine, with an activation energy of 17.9 kcal/mol in the presence of the deposit, showing product-catalyzed reaction similar to that of arsine [4]. Its use in MOCVD is straightforward. It produces layers with excellent surface morphology but poor electrical quality; namely, high p-type background carrier concentrations and low mobilities due to high carbon content [5]. Excessive carbon incorporation is a characteristic feature of the growth and has been shown unequivocally to come from the methyl groups of the TMAs molecule [6]. Thermal pre-cracking of TMAs, evidently to the dimer As_2, has been used to improve the electrical properties of GaAs grown by MOCVD [7], by vacuum chemical epitaxy [8], and by chemical beam epitaxy [9]; although the carbon content can be reduced by an order of magnitude, it is still far too high.

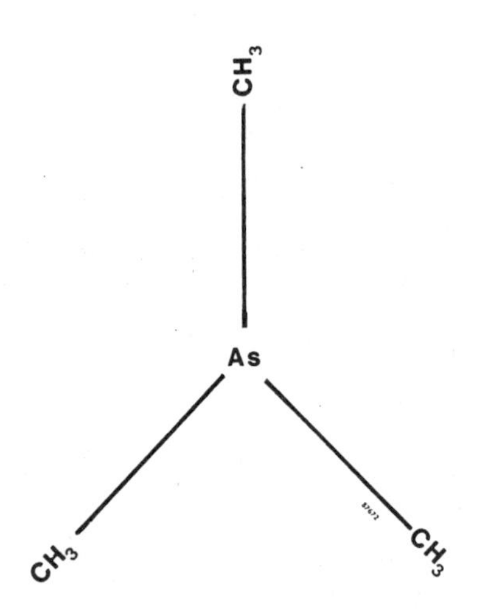

FIGURE 1. Structural formula of trimethylarsine (TMAs).

Triethylarsine

It was hoped that the use of TEAs (Figure 2) instead of TMAs would lead to the same pronounced reduction in electrically active carbon content as does the substitution of TEGa for TMGa in conventional growth with arsine [10]; TEAs was also expected to have a higher pyrolysis rate than TMAs. However, although the use of TEAs does yield a higher growth rate while maintaining good surface morphology, carbon contamination persists. MOCVD-grown GaAs was found to have high p-type background compensation and low carrier mobilities, with carbon levels up to the mid 10^{17} cm^{-3} range; in addition, a 10X buildup of carbon was observed in the interface region [11,12]. The carbon was attributed tentatively to the incorporation of both methyl radicals from the TMGa and incompletely pyrolyzed TEAs fragments [12].

Tertiarybutylarsine

TBAs (Figure 3) is a borderline-pyrophoric liquid of undetermined toxicity. It is expected to pyrolyze by C-As bond fission to yield highly reactive AsH_2 radicals and readily removed hydrocarbon radicals; thus, it should not contribute to carbon incorporation. The results to date are promising. The pyrolysis of TBAs is more

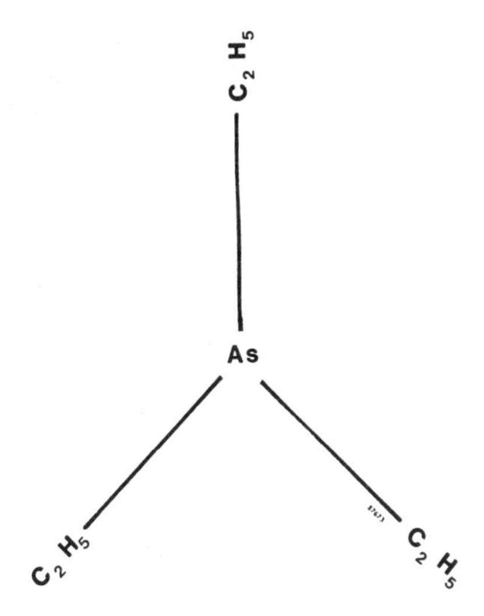

FIGURE 2. Structural formula of triethylarsine (TEAs).

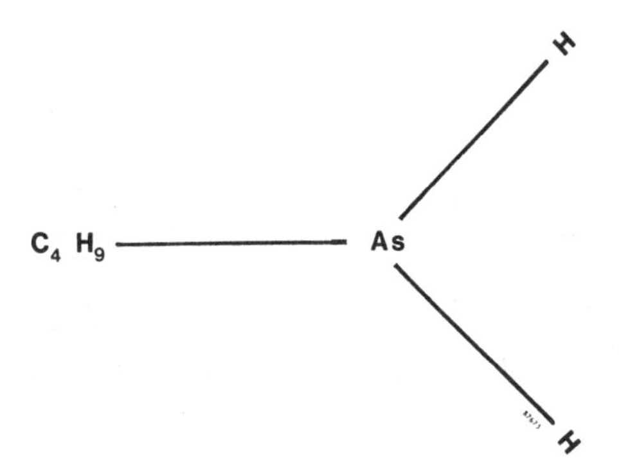

FIGURE 3. Structural formula of tertiarybutylarsine (TBAs).

rapid than that of arsine under similar conditions, so that excellent surface morphology is obtained at V/III ratios as low as 2 [13]. Deposited layers are always n-type, and carbon content is less than with TMAs, with the best material showing a background free-carrier concentration of 5×10^{15} cm^{-3} and a mobility at 27°C of 4000 cm^2/Vs [36].

Diethylarsine

DEAs (Figure 4) is an attractive alternative to arsine: it is a nonpyrophoric liquid with a toxicity that is presumed to be much lower. As with TBAs, the presence of As-bonded hydrogen atoms should facilitate the loss of the methyl groups from the decomposition of TMGa, and it should thereby minimize carbon incorporation. According to the single published study [15], the use of DEAs and TMGa yields high-quality GaAs layers with excellent surface morphology and low carbon content.

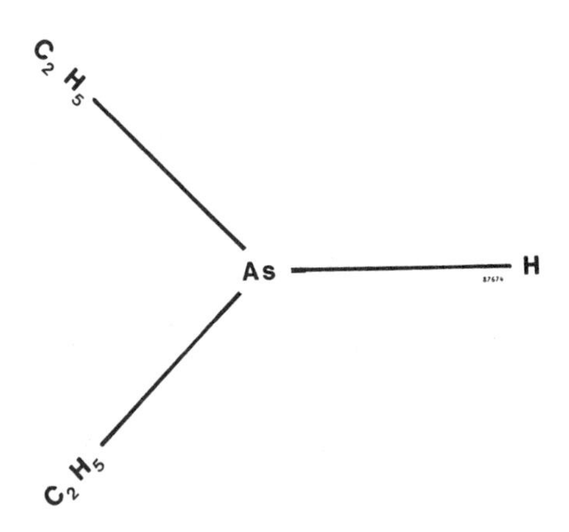

FIGURE 4. Structural formula of diethylarsine (DEAs).

The growth rate was found to be mass-transport-limited above 500°C and kinetically controlled at lower temperatures, with the best layers grown at 500-580°C. Indeed, these were the purest GaAs layers ever reported for MOCVD from TMGa and an As source other than arsine; the n-type background free-carrier concentration was $0.3\text{-}5 \times 10^{15} cm^{-3}$, and the 77°K mobility was 10,300-64,600 cm^2/Vs.

Dimethylarsine

DMAs (Figure 5) is a flammable liquid which is expected to closely resemble DEAs in its putative low toxicity and its freedom from excessive carbon incorporation. It is also expected to have a significantly higher vapor pressure than DEAs, which would facilitate its use as a source chemical for MOCVD. Although no data have been published, DMAs has been made available to several MOCVD specialists and is currently being evaluated as an alternative As precursor for GaAs growth.

SUMMARY

The few published studies of GaAs MOCVD with TMGa and liquid arsine sources are in agreement with the model of Kuech and Veuhoff [16] for carbon incorporation and indicate the need for proximate hydrogen

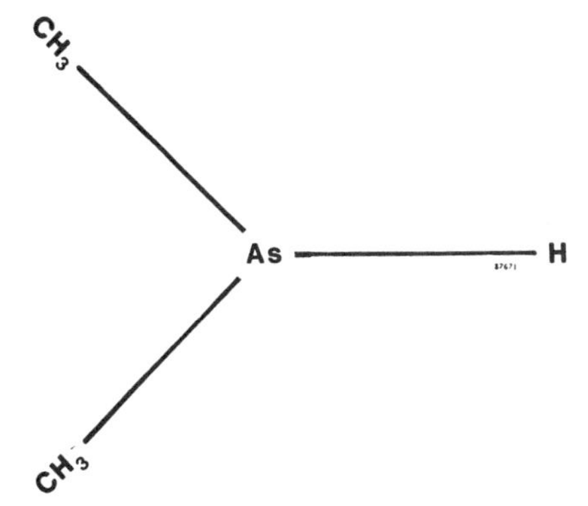

FIGURE 5. Structural formula of dimethylarsine (DMAs).

atoms to facilitate desorption of methyl radicals from the growth surface. On this basis, the fully substituted compounds TMAs and TEAs are not suitable alternatives to arsine. However, as suggested by Speckman and Wendt [12], TEAs might be used with TEGa in an exclusively ethyl-chemistry growth process, since both precursors decompose by beta-hydride elimination to yield non-reactive hydrocarbon byproducts [11, 17]. The reported data on the quality of GaAs grown by MOCVD with TMGa and TEAs, TBAs, or DEAs are presented in Table 2.

Four conclusions can be drawn from this necessarily scanty tabulation:

- o TBAs and DEAs both give low carbon incorporation; i.e., carbon levels comparable to those obtained with arsine.
- o TBAs and DEAs both undergo high-yield decomposition within the preferred deposition regime, permitting V/III ratios as low as unity.
- o The optimal growth temperatures with DEAs are at least 50C degrees lower than those with arsine.
- o For all three liquid sources, higher purity is required.

The importance of minimizing carbon incorporation by the selection of suitable precursors cannot be overstated; as pointed out by Keuch *et al.* [18], carbon is the sole intrinsic impurity limiting the electrical quality of MOCVD GaAs. In contrast, the extrinsic impurities which are introduced as contaminants in the liquid As sources can be reduced by higher-purity precursor production; an example is the germanium contamination of DEAs, which is held responsible for lowering the carrier mobility to about half the value obtained with arsine [15]. Lower V/III ratios provide major

TABLE 2. Quality of GaAs grown by MOCVD with TMGa and liquid As sources

	TEAs [11,12]	TBAs [13,14]	DEAs [15]
Growth Temperatures Investigated	540-700°C	530-800°C	380-700°C
V/III Ratios Investigated	6.7-40	1-20	0.9-5
Highest-quality Film Reported:			
Growth Temperature	650°C	700°C	550°C
V-III Ratio	40	2	0.9
Background Carrier Concentration, cm^{-3}	$2.2x10^{15}n$	$5 \times 10^{15}n$	$3 \times 10^{14}n$
Mobility, cm^2/Vs	10,000 @77°K	4,000 @300°K	64,600 @77°K 7,700 @300°K
Relative Carbon Level, from Low-temperature Photoluminescence	Excessive	Low	Low
Relative Exrinsic Impurity Level, from Low-temperature Photoluminescence and Secondary Ion Mass Spectrometry	Excessive	Excessive	Excessive

economies in reagent usage and, especially, effluent treatment. Lower growth temperatures are desirable for optimizing device structures.

In summation, both TBAs and DEAs appear to be attractive liquid sources of As. DMAs may also prove to be a viable alternative to arsine, especially since its vapor pressure is expected to be much higher than that of DEAs. In principle, any of the mono- and disubstituted alkyl derivatives of arsine hold promise. The final selection will require toxicity data, which can be obtained only by time-consuming studies on suitable animal populations.

REFERENCES

1. Wolfson, R.G. and Vernon, S.M., Safety in Thin Film Semiconductor Deposition, Solar Cells 19 (1986-87), 2156.

2. Bhat, R., OMCVD Growth of GaAs and AlGaAs Using a Solid As Source, J. Electron. Mater. 14 (1985), 433.

3. Cooper, C.B., III, et al., The Organometallic VPE Growth of $GaAs_{1-y}Sb_y$ Using Trimethylantimony and Ga_{1-} In As Using Trimethylarsenic, J. Electron. Mater. 9 (1980), 299.

4. Cherng, M.J., Cohen, R.M., and Stringfellow, G.B., $GaAs_{1-}$ Sb Growth by OMVPE, J. Electron. Mater. 13 (1984), 799.

5. Lee, P. et al., MOCVD in Inverted Stagnation Point Flow, I. Deposition of GaAs from TMAs and TMGa, J. Cryst. Growth 77 (1986), 120.

6. Lum, R.M., et al., Investigation of Carbon Incorporation in GaAs Using ^{13}C-enriched Trimethylarsenic and ^{13}C-methane, Electronic Materials Conference (Santa Barbara, CA, 1987), Paper E-3.

7. Vook, D.W., Reynolds, S., and Gibbons, J.F., Growth of GaAs by Metalorganic Chemical Vapor Deposition Using Thermally Decomposed Trimethylarsenic, Appl. Phys. Lett. 50 (1987), 1386.

8. Fraas, L.M., et al., GaAs Films Grown by Vacuum Chemical Epitaxy Using Thermally Precracked Trimethyl-arsenic, J. Appl. Phys. 62 (1987), 299.

9. Tsang, W.T., Chemical Beam Epitaxy of InP and GaAs, Appl. Phys. Lett. 45 (1984), 1234.

10. Kuech, T.F. and Potemski, R., Reduction of Background Doping in Metalorganic Vapor Phase Epitaxy of GaAs Using Triethylgallium at Low Reactor Pressures, Appl. Phys. Lett. 47 (1985), 821.

11. Hata, M., et al., OMVPE Growth of GaAs Using Triethylarsine, Electronic Materials Conference (Santa Barbara, CA, 1987), Paper E-4.

12. Speckman, D.M. and Wendt, S.P., Alternatives to Arsine: the Atmospheric Pressure Organometallic Chemical Vapor Deposition Growth of GaAs Using Triethylarsenic, Appl. Phys. Lett. 50 (1987), 676.

13. Chen, C.H., Larsen, C.A., and Stringfellow, G.B., Use of Tertiarybutylarsine for GaAs Growth, Appl. Phys. Let. 50 (1987), 218.

14. Lum, R.M., Klingert, J.K., and Lamont, M.G., Use of Tertiarybutylarsine in the Metalorganic Chemical Vapor Deposition Growth of GaAs, Appl. Phys. Lett. 50 (1987), 284.

15. Bhat, R., Koza, M.A., and Skromme, B.J., Growth of High-quality GaAs Using Trimethylgallium and Diethylarsine, Appl. Phys. Lett. 50 (1987), 1194.

16. Keuch, T.F. and Veuhoff, E., Mechanism of Carbon Incorporation in MOCVD GaAs, Second International Conference on Metal-organic Vapor Phase Epitaxy (Sheffield, England, 1984), Poster 31.

17. Yoshida, M. and Watanabe, H., Mass Spectrometric Study of $Ga(CH_3)_3$ and $Ga(C_2H_5)_3$ Decomposition Reaction in H_2 and N_2, J. Electrochem. Soc. 132 (1985), 677.

18. Kuech, T.F., et al., Properties of High-purity Al Ga_{1-} As Grown by the Metal-organic Vapor-phase-epitaxy Technique Using Methyl Precursors, J. Appl. Phys. 62 (1987), 633.

SAFER GROWTH OF GaAs USING TRIMETHYL ARSENIC*

Larry D. Partain and Robert E. Weiss
Varian Research Center, Palo Alto, CA 94303

Paul S. McLeod
Varian Associates, Santa Clara, CA 95054

ABSTRACT

Gas source growth of GaAs using arsine offers advantages for large scale fabrication of a range of devices including solar cells. The toxicity of arsine presents substantial problems of safety and liability. This provides a strong incentive to develop alternate gas sources for As. Trimethyl arsenic is a promising candidate that is up to 6500 times less toxic than arsine and is a low vapor pressure liquid that is much easier to contain and neutralize in case of an accident. It has been used in a vacuum MOCVD system to grow GaAs with a background p doping level of 6-8 x 10^{16} cm^{-3}, a mobility of 276 $cm^2/Vsec$, and room temperature photoluminescence full width, half height of 41 meV. This should be sufficient for the growth of high performance solar cells.

INTRODUCTION

GaAs devices offer performance advantages for photovoltaics, optoelectronics and high-speed circuits. The large scale production of AlGaAs/GaAs photocathodes for night vision systems began in 1982. Varian currently holds the world record for space solar cell efficiency at 21.5% using an AlGaAs/GaAs structure.[1] Both devices are made in a gas source metalorganic-chemical-vapor-deposition (MOCVD) system with arsine gas as the As source. The first large scale production (1000 2-cm x 2-cm cells per week) of high efficiency GaAs solar cells is from a MOCVD system using arsine.[2] Gas sources are the preferred approach to large scale fabrication because of their convenient control and scalability and their demonstrated capability.

Unfortunately, arsine is a hazardous material with potential safety and liability problems. It is lethal at concentrations below 100 ppm,[3,4] and it is a high pressure gas that spreads quickly in an accidental release. GaAs growths are made without arsine in halide transport gas source systems, but then Al-containing ternary compounds cannot be grown.[5] Such ternaries are required for the best solar cells and photocathodes, as well as for high performance lasers and high-electron-mobility-transistors (HEMTs).

Metalorganic (MO) gas sources of As offer two major advantages. First, they can be much less toxic than arsine; second, they are low vapor pressure liquids whose vapors evolve slowly in an accidental spill. This makes them much easier to contain and neutralize in an

---

*This work was performed while the authors were employed at Chevron Research Co., Richmond, CA.

accident. Here we report the use of the MO trimethyl arsenic (TMAs) to grow GaAs in a MOCVD system operated at vacuum (10^{-3} Torr) pressures.[6] We found that a thermal cracker is required for high quality material from this system, and that the cracker's temperature, pressure and catalyzing surfaces are the most important control parameters.

HAZARDS

Toxicology experiments have been reported for both arsine and TMAs where mice were subjected to lethal doses. For arsine exposure for 10 min, the lethal concentration at which 50% of the mice died (LC_{50} mice) was 78 ppm.[3,4] However, for TMAs exposure for 4 hours, the LC_{50} for mice was 21,300 ppm.[3,7] This specifies a factor of 273 higher concentration for TMAs for the same toxicity without allowing for the exposure time difference. There is at least one report that the product of lethal concentration and exposure period is a constant for arsine.[4] If this is accurate, the decreased toxicity factor for TMAs is 6550 compared to arsine.

TMAs is also safer because it is a liquid at room temperature, with a vapor pressure less than 1 atmosphere (230 Torr at 20°C).[8] The pure arsine typically used in vacuum MOCVD systems is a liquid at room temperature with a vapor pressure of 205 psig. For 1 atmosphere MOCVD, arsine is typically used as a gas diluted to 10% concentration in hydrogen, with a 1800 psig total pressure.

There has been recent speculation that arsine might spontaneously react with oxygen in air to form solid arsenic trioxide that is less hazardous. However, a comprehensive report[4] describes numerous experiments where arsine was mixed with air and used to determine LC_{50} values for 10-30 minute exposure times with mice, rats, rabbits and dogs. There is no indication that this air exposure decreased the toxicity below the values that are stated in these studies, and that are used to set present exposure limits.

EXPERIMENTAL RESULTS

A potential problem for GaAs grown in MOCVD reactors with TMAs is carbon incorporation from the incomplete decomposition of TMAs. However, this can be overcome with an upstream thermal cracker coupled with a MOCVD reactor operated at vacuum pressures so that decomposition products can be efficiently transported to the growth surface. Figure 1 shows the cracker used in such a system. We have given a more complete description of this elsewhere, and we call the system vacuum-chemical-epitaxy (VCE).[6,9]

The first line of Table I shows the transport properties obtained with no pinhole cap in the end of the cracker tube (described as 1/4-in i.d. cracker tube) and illustrated in Fig. 1. The background hole concentration was approximately 10^{19} cm^{-3} and equal to the carbon incorporation due to poor cracking of the TMAs. The hole concentration was measured with both van der Pauw Hall measurements and electrochemical $1/C^2$ versus V profiles. SIMS gave the carbon incorporation. The Hall mobility was a low 70 cm^2/V-sec, and the surface morphology was rough. The second line of Table I shows that the background hole concentration and the carbon incorporation decreased by an order of

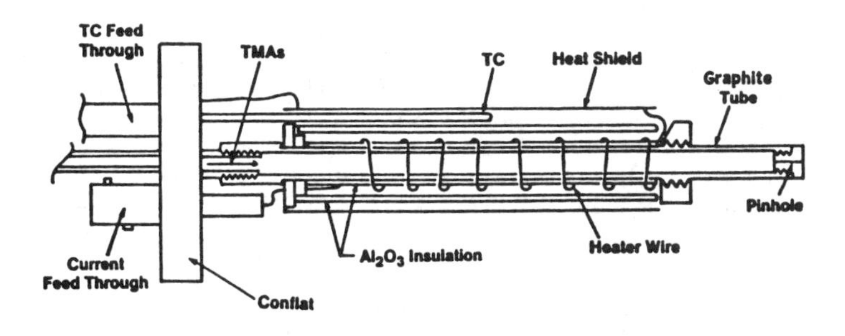

Fig. 1. Cracker furnace used for TMAs thermal decomposition.

magnitude when a cap was placed on the end of the cracker tube with a 20-mil diameter pinhole opening. However, the morphology remained rough, and the mobility improved although still suppressed in magnitude. The best properties came with the 13-mil pinhole filled with two 7-mil Ta wires. Then the background hole concentration reduced to the 6-8 x 10^{16} cm^{-3} level, the Hall mobility increased to 276 cm^2/V-sec, the morphology became smooth, and the carbon content decreased by another order of magnitude. Since GaAs solar cells are typically doped in the 10^{17}-10^{18} cm^{-3} levels, this material should give high performance photovoltaic devices without limits on carrier transport beyond that just due to the doping.[10]

Apparently the pinhole cap increased the pressure inside the cracker tube. This, and the catalytic action of the Ta, improved the

Table I. GaAs film properties with changes in the thermal cracker configuration.

Cracker Config.	Hall Parameters: Mobility (cm^2/Vsec.)	Hall Parameters: Hole Conc. (10^{16} cm^{-3})	Electrochem. Hole Conc. (10^{16} cm^{-3})	SIMS Carbon Conc. (10^{16} cm^{-3})
1/4-inch i.d.	70	1000	800	1500
20-mil pinhole	125	90	-	150
13-mil pinhole filled with Ta wire	276	6	8	10

cracking efficiency of the TMAs and decreased the background p-type carbon doping. Figure 2 shows that the cracker temperature was also critical for high quality growths. It plots the signal strength from the VCE's quadrupole mass spectrometer as a function of the cracker's thermocouple temperature. Above about 850°C, the TMAs signal decreased and the by-product ethane/ethylene and methane signals increased as the TMAs was efficiently decomposed. All the Table I data came with a 850°C thermocouple temperature.

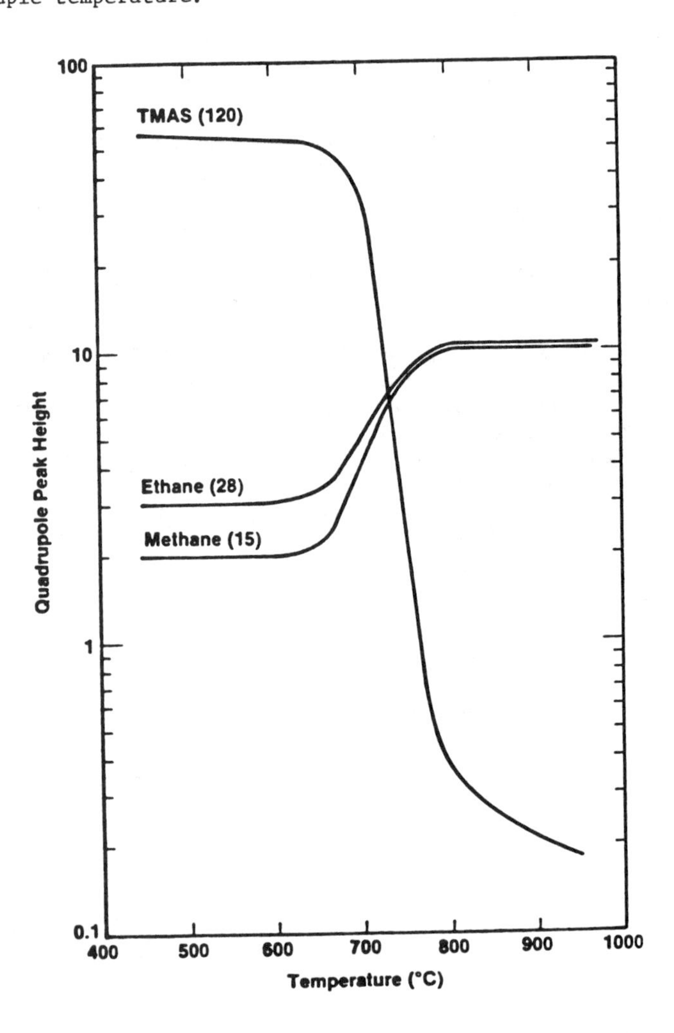

Fig. 2: TMAs (120 amu), methane (15 amu) and ethane/ethylene (28 amu) peak heights measured with a quadrupole mass analyzer as a function of the cracker thermocouple temperature.

Curve C of Fig. 3 shows the room temperature photoluminescence signal from one of our best VCE growths with TMAs. It has a full width, half height of 41 meV. This low value is an indication of its quality. For comparison, Curves B and A show the broader room temperature widths of 81 meV that Tsang[16] obtained with TMAs (and a thermal cracker in a vacuum MOCVD system that he calls chemical-beam-epitaxy, or CBE) and of 65 meV that he obtained on a 2 x $10^{18}cm^{-3}$ n-doped substrate used as a quality standard. Even better materials than these should come with further optimization of the three critical parameters we have identified: cracker temperature, pressure and catalyzing surfaces.

SUMMARY AND CONCLUSIONS

Gas source growth of GaAs offers potential advantages for the large scale fabrication of the highest performance solar cells, optoelectronic devices and high frequency components. So far, such fabrication has relied mainly on arsine as the gas source for As. However, arsine has serious safety and liability problems that provide a strong incentive to develop other less hazardous As sources. TMAs is up to 6500 times less toxic than arsine. In addition, it is a low vapor pressure liquid that is much easier to contain and neutralize in an accident than the high pressure arsine. We have used TMAs in a vacuum MOCVD system (called VCE) to grow GaAs with good background doping, mobility and photoluminescence properties that should be adequate for solar cell fabrication. An essential element for high quality material in a vacuum growth system is a thermal cracker for the TMAs. The most important cracker parameters are temperature, pressure and catalyzing surfaces.

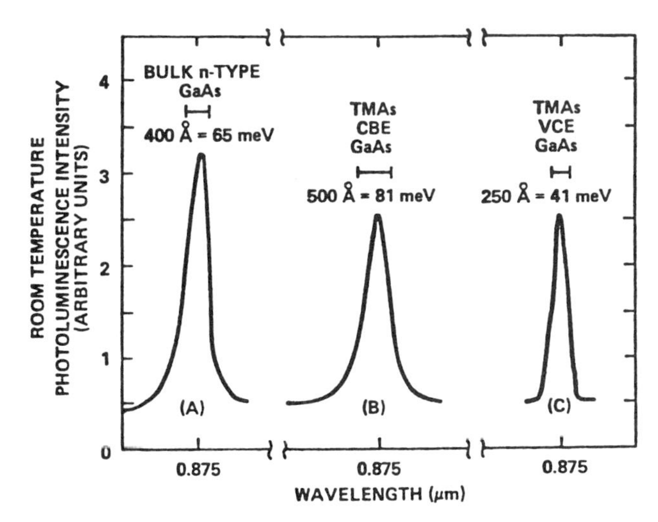

Fig. 3. Room temperature photoluminescence spectra measured on: (A) an n-type GaAs bulk substrate, and (B) GaAs grown with TMAs in a chemical-beam-epitaxy system. Curve C is the current work where GaAs was grown with TMAs in a vacuum-chemical-epitaxy system.

ACKNOWLEDGMENTS

Appreciation is due to Joan Murphy for obtaining a copy of the Government report cited. This work was supported in part by a research contract with the Solar Energy Research Institute.

REFERENCES

1. J. C. Werthen, G. F. Virshup, C. W. Ford, C. R. Lewis and H. C. Hamaker, Appl. Phys. Lett. 48, 1 (1986).
2. J. L. Tandon and Y. C. M. Yeh, J. Electrochem. Soc. 132, 662 (1985).
3. L. D. Partain, L. M. Fraas, P. S. McLeod and J. A. Cape, Solar Cells 19, 245 (1986-87).
4. Chemical Warfare Agents and Related Chemical Problems, Parts I-II, Vol. 1, Office of Scientific Research and Development, Contract No. OEM SR-1131, NTIS PB-158 508 (1946), p. 97.
5. G. B. Stringfellow, J. Crys. Growth 55, 42 (1981).
6. L. M. Fraas, P. S. McLeod, L. D. Partain and J. A. Cape, J. Appl. Phys. 62, 299 (1987).
7. C. Brekenridge, C. Collins, B. Hollomby and G. Lulham, Society of Toxicology Annual Meeting, Salt Lake City (1983), Abstract 93.
8. Alfa Products, Organometallics-Literature and Product Review (Morton Thiokol, Danvers, MA, 1986), p. 69.
9. L. M. Fraas, P. S. McLeod, L. D. Partain and J. A. Cape, J. Vac. Sci. Technol. B4, 22 (1986).
10. J. S. Blakemore, J. Appl. Phys. 53, R123 (1985).
11. W. T. Tsang, J. Electron. Mater. 15, 235 (1986).

REPLACEMENTS FOR ARSINE AND PHOSPHINE IN MOCVD PROCESSING

Donald Valentine, Jr.
American Cyanamid Company
1937 W. Main Street
Stamford, CT 06904

Many potentially hazardous chemicals are used in wafer processing for semiconductor manufacturing. Such hazardous production materials (HPMs) are potentially dangerous to humans because of flammability, corrosivity, toxicity, or reactivity, [1-6]. Binary hydrides, ExHy, such as arsine, phosphine, etc., have attracted much negative publicity due to life threatening effects (arsine, phosphine, diborane - the toxic hydrides), pyrophoric behavior (phosphine, silane), or corrosive properties (hydrogen chloride, ammonia).

The semiconductor industry response to the potential dangers involved in handling of HPMs has combined the use of often elaborate containment systems with sophisticated monitoring systems, [4, 7, 8]. These measures reduce but cannot eliminate the possibility of containment failures leading to large releases with substantial consequences in both the workplace and the surrounding environment.

As public concern grows, in spite of advanced containment and monitoring, increases are expected in the already high costs for acceptable controls. Use of hydrides such as arsine may be severely restricted or even forbidden in some locations. Availability and price may also become issues as manufacturers face increasing insurance costs and production regulations themselves.

In addition to the technical and political problems associated with the safe handling of phosphine, arsine and other binary hydrides, there are performance drawbacks to the use of these hydrides. Purity and batch-to-batch consistency have been improving but contamination with water, oxygen, and metals are recurring problems. More significantly, both arsine and phosphine have relatively high inherent thermal stabilities in non-oxidizing or reducing environments. High thermal stabilities translate into processing difficulties such as the need for higher than desirable growth temperatures in processes such as growth of III-V compound semiconductors by metal organic chemical vapor deposition (MOCVD). For example, in growths of gallium arsenide and indium phosphide the relatively high group V hydride stabilities require the use of wasteful high V/III reagent ratios and create difficulties in control of the growth processes.

Our MOCVD reagent program at American Cyanamid has explored whether new sources can offer better solutions to existing problems than continued efforts to work with existing materials. Our studies of new phosphorus and arsenic sources are briefly described in this paper.

POTENTIAL REPLACEMENTS FOR ARSINE AND PHOSPHINE

Arsine and phosphine based processes for fabrication of semiconductor structures include MOCVD of III-V thin films and processes such as ion implantation, CVD oxide processes and in-situ doping. Different arsine and phosphine replacement reagents may be appropriate depending on the application and, in particular, on whether arsenic (phosphorus) is oxidized (CVD oxides) or reduced (In-situ doping, ion implantation, MOCVD) in the process. Several arsine and phosphine replacements are available for the oxidative process - e.g. PCl_3, $POCl_3$, etc.

At American Cyanamid, our work has focused on development of arsine and phosphine replacements for use in reductive processes, especially for use in MOCVD processes. This work has led to identification of TBA and TBP as promising replacements for arsine and phosphine respectively.

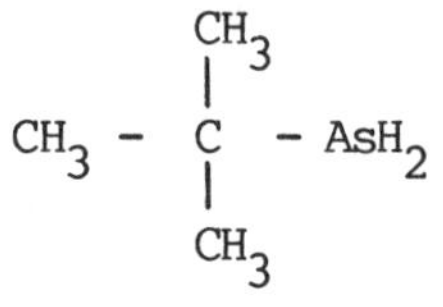

$$CH_3 - \underset{\displaystyle CH_3}{\overset{\displaystyle CH_3}{\underset{|}{\overset{|}{C}}}} - AsH_2 \qquad\qquad CH_3 - \underset{\displaystyle CH_3}{\overset{\displaystyle CH_3}{\underset{|}{\overset{|}{C}}}} - PH_2$$

Mono-tert.-butylarsine TBA   Mono-tert.-butylphosphine TBP

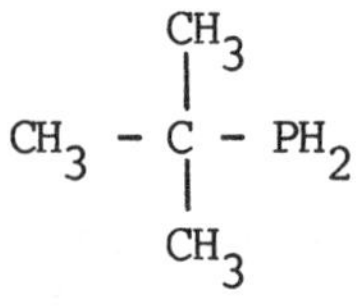

Advantages available from use of TBA and TBP comprise both improved safety with reduced potential environmental impact, and potentially improved performance in MOCVD, ion implantation, and in-situ doping. This paper is concerned with safety and MOCVD performance of TBA and TBP.

Table I provides a summary of physical and chemical properties of TBA and TBP, compared to the corresponding hydrides. Key features are:

- TBA and TBP are liquids with convenient vapor pressures at ambient temperatures, not compressed gases;

- TBA and TBP are pyrophoric according to some standard tests but are less readily ignited than PH_3 for example;

- TBA and TBP are less hazardous than parent hydrides. Toxicity studies were conducted on TBA and TBP at Bushy Run Research Center in Export, PA. For TBP, three hour exposures of rats to 100 - 1,144 ppm resulted in no mortalities indicating an LC50 of greater than 1,000 ppm, at least 20 times greater than for phosphine (approx. 10-50 ppm, apparently depending on purity). Toxicity data on TBA indicate a four hour LC50 of 70 ppm, indicating lower toxicity than arsine gas. The toxic mechanism for TBA has not been fully defined but appears to resemble more that for inorganic arsenic than for arsine. The liquid nature of both compounds also significantly reduces dispersion related hazards versus the use of the compressed hydride gases. Finally, TBA is rapidly oxidized in air, reducing the dispersion hazard. Modeling of several uncontrolled release scenarios suggests significant safety factors favoring either TBA or TBA over the corresponding hydrides. These are available on request. Users must decide relative safety factors based on their particular facility, however;

- TBA and TBP levels can be monitored using standard equipment. Calibration work is in progress and will be reported elsewhere.

Potential performance advantages of TBA and TBP are exciting. Both TBA and TBP appear to be able to solve performance problems encountered in previous attempts to replace arsine and phosphine by other sources for MOCVD.

Earlier studies with trimethylphosphine (TMP), [9], had revealed it was too unreactive to be useful as a phosphorus source in indium phosphide growth. Likewise, studies with trimethylarsine (TMAs), [10], and triethylarsine (TEAs), [11], have shown that gallium arsenide layers grown using TMAs or TEAs contained unacceptable carbon contamination from what appears to be an intrinsic growth mechanism of the tri-alkyl group V's. In addition, the purity of many currently available tri-alkyl compound derivatives is usually low. MOCVD growth results and general chemical studies indicate, however, that TBA and TBP are good replacements for the corresponding hydride gases. Both TBA and TBP decompose rapidly at lower temperatures than the corresponding hydrides and are more reactive in layer growth. Actual III-V growth rates using phosphine vs. TBP or arsine vs. TBA are the same at a given temperature, being limited by Group III reagent mass transport. The reactivities of TBA and TBP at lower temperatures can be used to advantage to develop superior growth processes and are attributed to the desirable properties of mono-alkyl group V derivatives containing the tert-butyl group.

Choice of the tertiarybutyl group was based on the expectation that it would perform better than other alkyls. First, both TBA and TBP are available using synthetic methods consistent with production of high purity materials. Second, the tertiarybutyl derivatives were expected to decompose at least partially by reactive fragment formation. Formation of reactive fragments AsH_2 and PH_2 occurs at lower temperatures than cracking of MH_3 and offers possibilities for performance improvements vs. MH_3. In situ hydride formation is undesired since the expected growth behavior would then be expected to be that of the hydride. Third, use of the tertiarybutyl group makes it possible to obtain a derivative with convenient volatility which can provide the desired inorganic MH_2 radical in a single M-C bond breaking step. This makes it easier to avoid C incorporation in the thin film.

$$(CH_3)_3C{-}MH_2 \xrightarrow{\text{IN SITU HYDRIDE FORMATION}} H_2C{=}C(CH_3){-}CH_3 + MH_3$$

ISOBUTYLENE

$$(CH_3)_3C{-}MH_2 \xrightarrow{\text{REACTIVE FRAGMENT FORMATION}} (CH_3)_3C\cdot + MH_2$$

TERTIARYBUTYL RADICAL

The thermal decomposition of the hydride replacements, TBP and TBA, is summarized in Table II, [12, 13]. It is evident that TBA pyrolyzes at significantly lower temperatures than AsH_3 with 100% decomposition at about 475°C vs. 760°C for the hydride. Likewise, TBP is 100% thermally decomposed at 550°C versus 850°C for PH_3.

The reduced decomposition temperature for TBA and TBP may provide significant advantages in growing various film systems at lower temperatures and reduced V:III ratios. There is some preliminary evidence of this using TBP in GaAsP growth (see Applications Section).

APPLICATIONS RESULTS

MOCVD applications results to date are very promising for TBA in gallium arsenide growth and for TBP in indium phosphide growth.

TBA

In a study comparing the MOCVD growth of gallium arsenide using trimethylgallium and arsine versus TBA, Chen, Larsen and Stringfellow found that TBA pyrolyzed at lower temperatures than did AsH_3, with the result that gallium arsenide layers with excellent morphology could be grown over a wider temperature range and at lower V/III ratios (approximately unity), [14]. Similar results were also found by Lum, [10]. Increasing purity of TBA has resulted in the growth of undoped GaAs mobility of 52,900 cm^2/Vs at American Cyanamid. Figure III shows photoluminescence spectra for two TBA runs and an AsH_3 baseline run immediately following the TBA growths. Note that spectrum B shows significant impurity reduction versus spectrum A. Unfortunately, the layer whose photoluminescence is shown as Spectrum B was too thin for full electrical characterization. The layer quality appears from the photoluminescence spectrum to be midway between the 52,900 cm^2/Vs (spectrum A) and the 110,000 cm^2/Vs (AsH_3 spectrum C).

Most TBA samples we have studied in MOCVD of GaAs give improving GaAs results as the bubbler is exhausted. This suggests that most of the electrically active impurities in the layers are derived from impurities remaining in the TBA. There appears to be no intrinsic barrier to growing high quality GaAs with TBA. Also, the fact that high purity GaAs was grown using AsH_3 immediately following TBA use indicates no risk of potential contamination from TBA in MOCVD systems. Studies are in progress of MOCVD of AlGaAs, p-GaAs and n-GaAs using TBA as the arsenic source. TBA purification remains under active study and higher purity TBA will soon be ready for study.

TBP

MOCVD growth of InP using TBP has resulted in indium phosphide layers with total carrier concentrations in low-mid 10^{15} cm^{-3} range, with room temperature mobilities of 4,000 cm^2/Vs, [15]. In many cases these properties are comparable to those achieved routinely with PH_3. In addition, G. B. Stringfellow at the University of Utah has explored the benefits of TBP in the growth of GaAsP, [16]. The substitution of TBP for PH_3 in this work resulted in a dramatic improvement in the phosphorus distribution coefficient and a nearly linear relationship between the solid and vapor compositions. The lower pyrolysis profile of TBP versus PH_3 gives a more temperature independent growth of mixed group V alloys (e.g. As and P), thereby potentially allowing for greater stoichiometry control in such material systems. This advantage is just beginning to be explored.

Contributions to this work were made by:

J. Baumann, D. Brown, H. Burkhard, C. J. Calbick, R. Fischer, G. Haacke, P. Kirlin, J. Kuninsky, D. Nucciarone, K. Olson, A. Robertson, M. Viscogliosi, S. Watkins.

Table I

Arsine and Phosphine vs. TBA and TBP

	AsH_3	TBA	PH_3	TBP
Formula	AsH_3	$C_4H_{11}As$	PH_3	$C_4H_{11}P$
Mol. Wt.	78	134	34	90
State (25°C)	Gas	Liquid	Gas	Liquid
BP (°C)	-55	69	-88	54
MP (°C)	-116	-1	-138	4
Vapor Pressure (Atm. at 25°C)	15	0.13	38	0.3
Purity (Metals)	Var.	6-9s	6-9s?	6-9s?
Purity (Oxygen)	Var.	1ppm	Var.	0.1ppm
Pyrophoric	No?	Yes*	Yes	Yes*
Inhalation Toxicity LC50 (ppm) 4h exposure	<25	75	11-50	>1100
Odor	Garlic?	Strong	Strong	Very Strong
Monitoring	Yes	Yes≠	Yes	Yes≠
Toxic Mechanism	Specific	As source	Specific	?

* TBA and TBP will pass some standard pyrophoricity tests but not others. Both should be handled as pyrophorics.

≠ Details on request.

Table II

Thermal Decomposition Profiles

Temperature °C

% Decomposition	AsH_3[13]	TBA[12]	PH_3[17]	TBP[17]
25	650	280	600	420
50	680	300	700	450
75	700	350	800	475
100	760	475	850	550

REFERENCES

1. Uniform Fire Code Standard, No. 79-3.

2. Bolmen, Jr., R.A., Semiconductor International, 5, (1986), 156.

3. Bolmen, Jr., R.A., Semiconductor International, 5, (1986), 231.

4. Burggraaf, P., Semiconductor International, 11, (1987), 55-62.

5. Wald, P.H., State of the Art Reviews: Occupational Medicine, 1(1), (1986), 105.

6. LaDou, J., Technology Review, 5, (1984), 23.

7. Henon, B.K., Semiconductor International, 4, (1987), 164.

8. Murray, C., Semiconductor International, 8, (1986), 50.

9. Renz, H.; Weidlein, J.; Benz, K.W.; Pilkuhn, M.H., Electronic Letters, 16(6), (1980), 228.

10. Lum, R.M.; Klingert, J.K.; Lamont, M.G., Appl. Phys. Letters, 50(5), (1987), 284.

11. Speckman, D.M.; Wendt, J.P., Appl. Phys. Letters, 50(11), (1987), 676.

12. Lee, P.W.; Omstead, T.R.; McKenna, D.R.; Jensen, K.F., ACGIH Meeting, Monterey, CA (July 1987).

13. Nishizawa, J.; Kurabayashi, T., J. Electrochem. Soc., 130(2), (1983), 413.

14. Chen, C.H.; Larsen, C.A.; Stringfellow, G.B., Appl. Phys. Letters, 50(4), (1987), 218.

15. Chen, C.H.; Cao, D.S.; Stringfellow, G.B., Appl. Phys. Letters, in press.

16. Chen, C.H.; Cao, D.S.; Stringfellow, G.B., OMVPE Workshop, Cape Cod (1987).

17. Chen, C.H.; Larsen, C.A.; Stringfellow, G.B.; Brown D.W.; Robertson, A.J., J. Crystal Growth, 77(11), 1986.

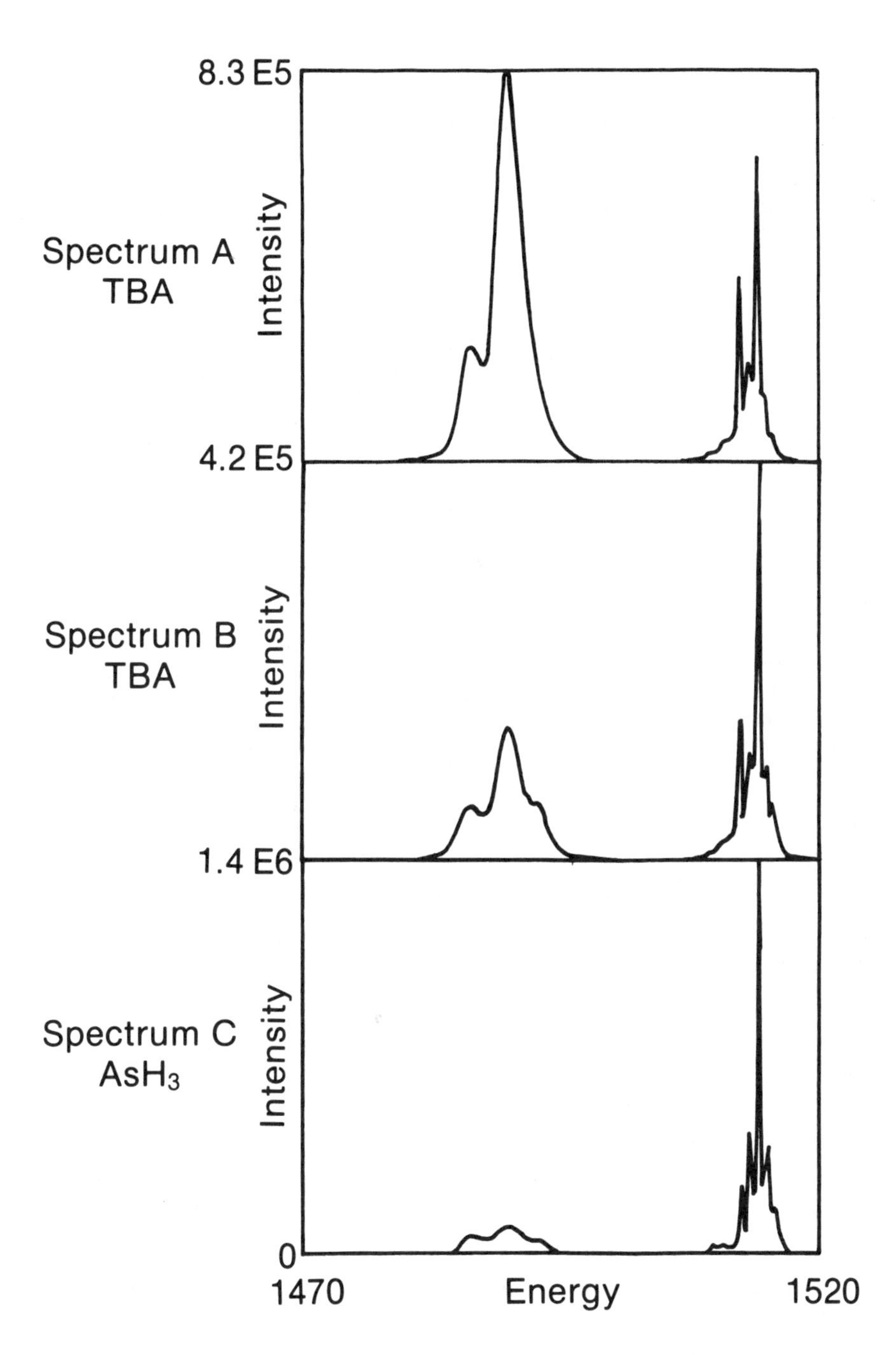

Figure III. Photoluminescence spectra for two TBA runs and one AsH_3 baseline run.

EXPOSURE AND CONTROL ASSESSMENT OF SEMICONDUCTOR MANUFACTURING

James H. Jones
National Institute for Occupational Safety and Health
4676 Columbia Parkway, Cincinnati, Ohio 45226

ABSTRACT

From 1980 to 1984, the National Institute for Occupational Safety and Health (NIOSH), along with the U.S. Environmental Protection Agency (EPA), sponsored a study of worker exposures and controls in semiconductor manufacturing. The study was conducted by Battelle Columbus Laboratories and PEDCO Environmental. Walk-through surveys were conducted at 21 plants and in-depth studies were done at four of these plants. Processes studied included photolithography, chemical vapor deposition, wet chemical etching and cleaning, plasma etching, diffusion, ion implantation, and metallization. Air samples were collected for acetone, antimony, arsenic, boron, n-butyl acetate, diborane, 2-ethoxyethyl acetate, hexamethyldisilizane, hydrogen chloride, hydrogen fluoride, 2-methoxyethanol, methyl ethyl ketone, nitric acid, phosphorus, sulfuric acid, and xylene. In addition, radio-frequency and ionizing radiation were monitored and ventilation measured. In general, results were well below recommended standards for routine operations. One exception was radio-frequency radiation where there was the potential for overexposure in several instances. Worker exposures during maintenance operations and process upset conditions were not able to be evaluated.

INTRODUCTION

Researchers from the National Institute for Occupational Safety and Health's (NIOSH) Engineering Control Technology Branch (ECTB) began evaluating worker exposure to toxic agents and the control of these exposures in semiconductor manufacturing in 1980. NIOSH at that time signed an interagency agreement with the Environmental Protection Agency (EPA) which included joint research in the electronics industry. Battelle Columbus Laboratories, under contract to EPA, in conjunction with ECTB, conducted a study of worker exposure and hazard control technology in the semiconductor industry. The study consisted of one-day walk-through surveys at 21 plants and week long in-depth surveys at four of these plants. During the in-depth surveys air samples were collected and measurements of physical agents made to determine worker exposures to various toxic agents.[1-4]

A number of processes were evaluated including photolithography, chemical vapor deposition, wet etching and cleaning, plasma etching, diffusion, ion implantation, and metallization. The agents evaluated for each process are shown in Table I.

Table I Hazardous agents evaluated

Process	Agent
Photolithography	acetone, n-butyl acetate, 2-ethoxyethyl acetate, hexamethyldisilizane, 2-methoxyethanol, methyl ethyl ketone, xylene
Chemical vapor deposition	hydrogen chloride
Wet etching and cleaning	hydrogen fluoride, nitric acid, sulfuric acid,
Plasma etching	radio-frequency radiation
Diffusion	antimony, arsenic, boron, phosphorus
Ion implantation	arsenic, boron, ionizing radiation, phosphorus
Metallization	radio-frequency radiation

METHODOLOGY

During the in-depth surveys, both personal and area air samples were collected to help characterize worker exposure and adequacy of controls. Air samples were collected for one to six hours and were used to estimate 8-hour time weighted averages (TWA's). The sampling and analytical methods used for each chemical are indicated in Table II.

Table II Sampling and analytical methods

Analyte	Method
Acetone	Adsorption on charcoal, desorption with CS_2, and analysis by mass spectroscopy with single ion monitoring
Antimony	Collection on 0.8-um mixed cellulose ester membrane filters, digestion with nitric acid, and analysis by flame atomic absorption spectroscopy (NIOSH P&CAM 73)[5]

Table II Sampling and analytical methods (cont.)

Analyte	Method
Arsenic, Particulate	See Antimony
Gas or Vapor	Adsorption on charcoal, desorption with nitric acid, and analysis by flameless atomic absorption spectroscopy (NIOSH S229)[5]
Boron	Adsorption on oxidizer-impregnated charcoal, desorption with 3% hydrogen peroxide, and analysis by plasma emission spectroscopy (NIOSH P&CAM 341)[5]
n-butyl acetate	See Acetone
2-ethoxyethyl acetate	See Acetone
Hexamethyldisilizane	See Acetone
Hydrogen chloride	Collection in 0.5 M sodium acetate, analysis by chloride ion-specific electrode (NIOSH S246)[5]
Hydrogen fluoride	Collection in 0.1 N sodium hydroxide, analysis by ion-specific electrode (NIOSH S176)[5]
2-methoxyethanol	See Acetone
Methyl ethyl ketone	See Acetone
Nitric acid	Collection in distilled water, analyzed by ion-specific electrode (NIOSH S319)[5]
Phosphorus	Adsorption on treated silica gel, desorption with hot acidic permanganate solution, formation of phosphomolybdate complex, extraction with isobutanol/toluene mixture, reduction with stannous chloride, and analysis by spectroscopy (NIOSH S332)[5]

Table II Sampling and analytical methods (cont.)

Analyte	Method
Sulfuric Acid	Collection on 0.8-um mixed cellulose ester membrane filter, extraction with isopropyl alcohol/ water, and titrated with 0.005M barium perchlorate (NIOSH S174)[5]
Xylene	See Acetone

Radio-frequency radiation was evaluated using a Holaday Model HI 3002 meter with an electric field probe having a frequency response of 500 kHz to 6 GHz and a magnetic field probe with a frequency response of 5 to 300 MHz. Readings were taken during normal process equipment operating cycles. Normal operating frequencies, power outputs, and cycle times were recorded. The results are reported for near-field measurements as maximum electric field strength (V^2/m^2) and maximum magnetic field strength (A^2/m^2).

Ionizing radiation at various work stations was surveyed with a either a Nuclear Chicago Model 2650 or Victoreen Model 4040 RFC Geiger-Muller counter. The magnitude of potential exposure at "hot" survey locations was characterized by using Landauer Type P-1 dosimetry badges. The Geiger-Muller survey results are reported in milliroentgens per hour (mR/hr), and the dosimeter readings are reported in millirems (mrem) for the the specific time period monitored.

RESULTS

Photolithography

Photolithography was performed predominately using automated Wafer Trak systems. Controls utilized were the partial enclosure of aerosol and vapor generating operations such as spin on application of primers, photoresists and developers. Local exhaust ventilation was applied to these enclosures and to the machine cabinet in general because of the reservoirs of process chemicals located there. General ventilation was also used, but primarily to keep particulates from contaminating the wafers. Consequently, air flow patterns were often not conducive to reducing worker exposure to chemicals. Air sampling was done at seven photolithography lines in three plants. Air monitoring results of the chemicals found in this process, primarily solvents, indicate that exposures during normal operations are very low. In most cases the results are far below OSHA Permissible Exposure Limits (PEL's) or NIOSH Recommended Exposure Limits (REL's) for the various chemicals. Hexamethyldisilizane has no standard or recommended guidelines. The sample results are presented in Table III.

Table III Photolithography air sampling results

Agent	No. Samp.	Geom. Mean	Concentration (ppm) Min.	Max.	OSHA PEL[6]	NIOSH REL[7]
Acetone	31	0.069	<0.002	3.7	1000	250
n-butyl acetate	31	0.004	<0.001	0.07	150	-
2-ethoxyethyl acetate	31	0.005	<0.002	0.52	100	*
Hexamethyldisilizane	31	<0.041	<0.013	0.23	-	-
2-methoxyethanol	31	<0.003	<0.001	<0.006	25	*
Methyl ethyl ketone	4	<0.013	<0.010	<0.017	200	200
Xylene	31	0.022	<0.001	4.9	100	100

* lowest level possible

Chemical Vapor Deposition

A low pressure chemical vapor deposition (LPCVD) process was evaluated at one plant for the decomposition product hydrogen chloride. The LPCVD cabinet had a scavenger box surrounding the load end of the furnace. Air was exhausted at the rate of 60 cfm from the scavenger box which created a capture velocity of 50 fpm with the scavenger box door open. Air sample results are reported in Table IV.

Table IV Chemical vapor deposition air sampling results

Agent	No. Samp.	Geom. Mean	Concentration (ppm) Min.	Max.	OSHA PEL[6]	NIOSH REL[7]
Hydrogen chloride	4	<1.9	<1.8	<1.9	5	-

Wet Etching and Cleaning

The wet etching and cleaning process is carried out in a polypropylene or stainless steel lab hood-like apparatus. The work surface has liquid wells that contain the etching and cleaning chemicals as well as deionized water rinse bathes. The work surface has a hood enclosure on three sides with local exhaust ventilation at the rear wall and sometimes also the work surface or around the liquid wells. A complicating factor often found is the use of a downdraft laminar air flow over the work surface to prevent particulate contamination of wafers. This added air flow into the hood means that there is a net flow of air out of the hood, so there is the possibility to entrain chemicals and carry them into the breathing zone of the worker. Wet benches at three plants were monitored. Most of the air sampling results were below detection limits. The only exception was one sulfuric acid sample, which was still below recommended standards. The results are reported in Table V.

Table V Wet etching and cleaning air sampling results

			Concentration (mg/m^3)			
Agent	No. Samp.	Geom. Mean	Min.	Max.	OSHA PEL[6]	NIOSH REL[7]
Hydrogen fluoride	14	<2.3	<0.9	<4	2.5	2.5
Nitric acid	2	<1.5	<1.5	<1.5	5	5
Sulfuric Acid	2	0.24	<0.081	0.74	1	1

Plasma Etching

Plasma etching is carried out in a vacuum chamber where etchant gases are released and converted by radio-frequency (RF) energy to a plasma. Workers spend little time at the etcher except to load and unload wafers. At this time the etcher is not generating RF energy. When operating, RF energy is prevented from being released into the work area by metal shielding and grounding of the etcher. Twenty three units from four plants were monitored for release of RF energy. All units evaluated were designed to operate at 13.56 MHz. Five of the 23 units evaluated emitted electrical field energies and three emitted magnetic field energies above the ACGIH Threshold Limit Value (TLV) for RF energy. Five of the six units exhibiting high leakage rates were older units with no shielding at the viewing port. Two of these units also had external recorders attached that were not properly grounded and were emitting RF energy. The energy levels reported below are maximum levels for each unit measured at 10 cm from the unit. The ACGIH TLV applies to any six minute period and is intended for whole body radiation. Few of the levels were at high enough levels far enough from the units to provide this whole body radiation. In addition, the workers were usually not near the units for a continuous time of six minutes during the time the RF energy is utilized. However, because the potential for localized high exposures exists and the RF energy is relatively easy and inexpensive to control, it seems prudent to eliminate these sources with proper grounding of attachments and the installation of viewing port shielding kits available from the unit manufacturers. Also, when maintenance is conducted on these units it is very important that shielding be reinstalled properly and that all missing screws be replaced. Monitoring results are presented in Table VI.

Table VI Plasma etching monitoring results

Agent	No. Samp.	Geom. Mean	Min.	Max.	OSHA PEL[6]	ACGIH TLV[8]
Radio-frequency radiation						
Electric field V^2/m^2	23	780	<200	400000	37700	18500
Magnetic field A^2/m^2	23	0.04	<0.02	0.37	0.265	0.13

Diffusion

Diffusion is carried out in furnace tubes similar to those used in chemical vapor deposition. Again, the primary controls are enclosure and local exhaust ventilation. The source cabinets, where doping agents are supplied to the furnaces, are ventilated, scavenger boxes enclosing the load end of the furnace tube are also ventilated, as is the furnace cabinet. One bank of diffusion furnaces using solid antimony trioxide as a doping source was monitored and no detectable levels were found. Another bank of furnaces using arsine as a dopant was monitored for decomposition products, arsenic trioxide, that might escape from the scavenger boxes. All results were below detection limits. Two banks of furnaces using diborane as a dopant were monitored. Levels found averaged about one half the OSHA PEL although none exceeded the PEL. Two banks of furnaces using phosphorus oxychloride as a dopant source were monitored for phosphorus, since decomposition products could not be speciated. All but one sample were below detection limits. Sample results are reported in Table VII.

Table VII Diffusion air sampling results

		Concentration (ug/m^3)				
Agent	No. Samp.	Geom. Mean	Min.	Max.	OSHA PEL[6]	NIOSH REL[7]
Antimony	2	<12	<11	<12	500	500
Arsenic	2	<0.05	<0.04	<0.05	10	2
Diborane	2	58	40	83	100	-
Phosphorus	7	<48	<13	79	-	-

Ion Implantation

Ion implantation is carried out in a vacuum chamber. In the process a dopant material is ionized, accelerated toward a targeted wafer, and then implanted into selected areas of the wafer. Several potential hazards are encountered in this process: toxic dopant materials, high electrical energy, and ionizing radiation. Controls consist primarily of enclosures local exhaust ventilation, and shielding. Six implanters using arsine as a dopant were monitored for arsine with the results reported as total arsenic. All samples were below the OSHA PEL, but one sample was above the NIOSH REL for arsenic. Three implanters using boron trifluoride as a dopant were monitored for boron, since it is not possible to speciate for boron trifluoride or its decomposition products. If the highest concentration found was all boron trifluoride, which is unlikely, it would still be well below the OSHA PEL. In addition, two implanters using phosphine as a dopant were monitored for total phosphorus since it was not possible to speciate for different decomposition products. All samples were below the limit of detection. Ionizing radiation was monitored at four of these implanters at two plants. Levels found were unlikely to produce worker exposures above the OSHA PEL of 1.25 rems per calendar quarter. Monitoring results are presented in Table VIII.

Table VIII Ion implantation monitoring results

Agent	No. Samp.	Concentration (ug/m^3) Geom. Mean	Min.	Max.	OSHA PEL[6]	NIOSH REL[7]
Arsenic - end station	25	0.49	<0.3	3.8	10	2
- source	8	3.5	<1.2	87	10	2
Boron - end station	10	0.53	<0.21	27	-	-
- source	1	7.5	-	-	-	-
Phosphorus - end station	4	<52	<20	<95	-	-
		Total emissions (mrems)			Avg. Weekly Dose (mrems)	
Ionizing radiation -end station	3	<10	<10	-	<1	
-source inside cabinet	4	236	<10	14850	1768	
outside cabinet	4	16	<10	70	2	

Metallization

Metallization is another process carried out in a vacuum chamber. A metal source is heated to produce a vapor which is then condensed on the wafer. In some cases the heat source is RF energy. Nine units in two plants were monitored. Levels measured at one unit were above the ACGIH TLV for electric field energy and three units were above the TLV for magnetic field energy. As in the plasma etcher results, these are the maximum levels measured at 10 cm from each unit and do not reflect whole body exposure. Workers do not generally spend more than six minutes continuously at the units while the RF energy is being produced. However it is prudent to try to minimize the RF leaks that are occurring. The emissions encountered with these units seemed to be the result of incorrect replacement of shielding and missing screws after maintenance procedures. Monitoring results are reported in Table IX.

Table IX Metallization monitoring results

Agent	No. Samp.	Geom. Mean	Min.	Max.	OSHA PEL[6]	ACGIH TLV[8]
Radio-frequency radiation						
Electric field V^2/m^2	9	890	<200	62000	37700	18500
Magnetic field A^2/m^2	8	0.17	<0.02	25	0.265	0.13

CONCLUSIONS

Based on the results of this study, worker exposures during routine production work appear to be controlled below existing standards or recommended standards. However, there were several situations where special attention needs to be given, such as, assuring that RF emissions are reduced as much as possible to eliminate the potential for overexposure. Attention needs to be directed at continuous low level arsenic exposures in ion implantation operations. Continuous monitoring is typically done to detect arsine leaks during operations, but alarm levels are concerned with preventing acutely toxic levels of arsine. Alarm set points are generally set at 25 to 50 ppb (100-200 ug/m^3) which is 50 to 100 times higher than the NIOSH REL which is based on carcinogenic effects of arsenic. Routine exposures above the NIOSH REL could occur, but because the continuous monitors give no response at this level, a plant might assume that no arsenic exposure was occurring. Exposures during maintenance or process upset conditions could not be evaluated. These conditions have a higher potential for worker overexposure than normal operating conditions, so special attention should be given to minimizing the potential for unexpected releases of toxic materials and to assuring worker protection during maintenance when normal controls may be inoperable. Also, materials are being used in the industry that have little or no toxicological information, so it is difficult to determine what safe levels of worker exposure to these materials are. Several materials are difficult or impossible to monitor with readily available techniques, such as being able to speciate different boron or phosphorus compounds. This is important because the different compounds have widely varying toxicity.

REFERENCES

1. L.J. Ungers, G.J. Mihlan, and J.H. Jones, In-depth Survey Report: Control Technology for the Microelectronics Industry Report No. CT-115-12b (DHHS,PHS,CDC,NIOSH, Cincinnati, Ohio, 1984).
2. L.J. Ungers, R.K. Smith, and J.H. Jones, In-depth Survey Report: Control Technology for the Microelectronics Industry Report No. CT-115-16b (DHHS,PHS,CDC,NIOSH, Cincinnati, Ohio, 1984).
3. L.J. Ungers, G.J. Mihlan, and J.H. Jones, In-depth Survey Report: Control Technology for the Microelectronics Industry Report No. CT-115-20b (DHHS,PHS,CDC,NIOSH, Cincinnati, Ohio, 1984).
4. L.J. Ungers, R.I. Mitchell, and J.H. Jones, In-depth Survey Report: Control Technology for the Microelectronics Industry Report No. CT-115-24b (DHHS,PHS,CDC,NIOSH, Cincinnati, Ohio, 1984).
5. National Institute for Occupational Safety and Health, NIOSH Manual of Analytical Methods, 2nd ed. DHHS (NIOSH) Publication No. 82-100 (DHHS,PHS,CDC,NIOSH, Cincinnati, Ohio, 1983).
6. Occupational Safety and Health Administration, Occupational Safety and Health Standard for General Industry 29 CFR Part 1910 (DOL,OSHA, Washington, D.C., 1979).
7. National Institute for Occupational Safety and Health, Pocket Guide to Chemical Hazards DHHS (NIOSH) Publication No. 85-114 (DHHS,PHS,CDC,NIOSH, Cincinnati, Ohio, 1985).
8. American Conference of Governmental Industrial Hygienists, TLV's Threshold Limit Values and Biological Exposure Indices for 1987-1988. (ACGIH, Cincinnati, Ohio, 1987).

TOXICOLOGY OF TETRAMETHYLTIN AND OTHER ORGANOMETALS USED IN PHOTOVOLTAIC CELL MANUFACTURE

L.D. Hamilton, W.H. Medeiros, P.D. Moskowitz, and K. Rybicka
Biomedical and Environmental Assessment Division
Brookhaven National Laboratory
Upton, New York 11973

ABSTRACT

In photovoltaic cell fabrication, organometals (alkyl metals) may be used in such processes as metalorganic chemical vapor deposition, transparent contact oxide deposition, doping, and ion implantation. Although these compounds offer potential performance advantages over earth metals and possibly greater safety in handling than metal hydrides, they are not without risk to health and property. Most organometals can ignite spontaneously in air. Some also react violently with water. Oxidation by-products from these reactions are hazardous to health. Of the organometals used in photovoltaic cell fabrication, only the toxicology of organotins (triethyl-, trimethyl- and tetramethyltin) was studied extensively. In mammalian systems, tetramethyltin is rapidly dealkylated to trimethyltin. Although tin was classified by some investigators as an essential trace element, the effects of organotin compounds on humans are poorly known. Animal studies show that the most prominent effects of trimethyltin are on the central nervous system. Several observations of poisoning were reported; effects ranged from reversible neurologic disorders to death. Limited available data suggest that humans respond to single acute doses and more alarmingly to repeated sub-toxic doses, suggesting a cumulative effect. Toxicologic properties of diethyltelluride also were evaluated in animal experiments. The compound had toxic effects on the blood, liver, kidney, heart, and skin. Based on these studies and others of related compounds (e.g., methylmercury, tributyltin) extreme caution should be exercised in using organometal compounds in photovoltaic cell manufacturing.

INTRODUCTION

The U.S. Department of Energy-National Photovoltaics Program (DOE-NPP) is sponsoring research and development on photovoltaic energy systems to identify critical technical barriers to improved efficiency and lower cost, and to increased reliability and durability of materials, devices, modules, and systems. In developing these alternatives, DOE is interested in ensuring that public and occupational health and safety are not endangered. In response, DOE-NPP is sponsoring research at Brookhaven National Laboratory (BNL) to identify potential health and safety hazards of selected material and process alternatives so that mitigation strategies can be incorporated into system designs before large-scale commercialization. As part of this effort, this paper summarizes a review by Rybicka et al.[1] of the physical and biological hazards of tetramethyltin and other organometals used in the photovoltaics industry, and the extensive reviews by McMillan and Wenger[2] and Herron.[3]

MATERIAL AND PROCESS OPTIONS

Some photovoltaic cell fabrication process options (e.g., vapor phase epitaxy) use large quantities of earth metals (e.g., aluminum and gallium) and binary element hydrides (e.g., arsine and hydrogen selenide). The solid earth metals are not unusually toxic or hazardous, but substitute materials are being explored because of poor performance characteristics in film deposition. In contrast, the hydride gases are highly toxic and hazardous (i.e., flammable or explosive) when used under elevated temperature and pressure. Hence, their use can present substantial risks to life and property. Because of the relatively poor performance of earth metals and the risks of binary element hydrides, organic substitutes with improved performance characteristics or reduced hazards are being developed or used in such process options as chemical vapor deposition oxide formation, doping, ion implantation, and metalorganic chemical vapor deposition (MOCVD) (Table I). Although these substitutes can offer potential performance advantages, or greater safety to the extent that they substitute liquids for gases, their use is not without risk. In this context, the following sections examine the inherent physical (flammability and explosivity) and biological (toxicity) hazards associated with their use.

CHEMICAL AND PHYSICAL PROPERTIES

The chemical and physical properties of organometals used in the photovoltaic industry are summarized in Table II. These organometals are colorless or pale liquids; the only exception is trimethylindium, which is crystalline at room temperature. Some have characteristic garlic (diethyl telluride) or musty (tetramethyltin) odors. They are highly reactive with air and some react violently with water. Hazardous properties of these materials are summarized in Table III.

BIOLOGICAL EFFECTS OF EXPOSURE

To gather data on potential biological effects of these materials, detailed computerized literature searches were done using the National Library of Medicine's Toxline, Medline, Registry of Toxic Effects of Chemical Substances and Toxicology Data Bank; and Dialog's CA Search, National Technical Information Service, and the Federal Register Abstracts. This search revealed that the toxicology of only the organotins (triethyl-, trimethyl-, and tetramethyltin) were studied extensively. Organotins were used in the manufacture of plastics, in agriculture as pesticides, in paint as fungicides and in glass manufacturing as a surface coating.

Tetramethyltin

Numerous studies have identified the form and degrees of organotin toxicity. The proliferation of recent research on organotins is more attributable to their use in the study of brain function than to their role in environmental toxicology. Most experimental studies focused on triethyltin (TET) and trimethyltin (TMT); only a few were devoted to tetramethyltin (TTMT). It is, however, generally recognized that TTMT is dealkylated in the body into TMT. Thus, it is assumed that both compounds produce the same effect. Therefore the studies on TTMT and TMT will be reviewed together.

EFFECTS ON ANIMALS

Absorption, Distribution, and Metabolism

The absorption, distribution, and metabolism of TTMT is based on its physical and chemical properties, which are analogous to those of other more intensely studied alkyltin compounds, notably tetraethyltin.[3] In mammalian systems TTMT is quickly transported to the liver where it undergoes rapid dealkylation to TMT.[4-6] There are differences in the rank order and tissue concentrations among different animals. In mice after intraperitoneal administration of TMT at 4.26 mg/kg of body weight, a peak in tissue levels was observed at 1 h in the kidney, liver, blood, lung, and testes. However, in the brain, skeletal muscle, and adipose tissue, a peak was not observed until 16 h after administration.[7] In contrast to mice, rats showed higher levels of tin in blood than in brain tissue at all experimental doses and time periods.[8]

Acute Toxicity

The toxic dose of TMT varies by species, strain, and route of administration.[8,9-13] The lethal dose in different animals was related to the binding of TMT by hemoglobin.[14] A relatively low lethal dose (LD_{50} approximately 3 mg TMT/kg of body weight[14]) was found for mice, hamsters, gerbils and marmosets in which TMT does not bind to hemoglobin. In contrast a high lethal dose is characteristic for rats (LD_{50} from 7.45 - 12.6 mg/kg of body weight) in which TMT binds to hemoglobin; this accounts for the high concentration in blood.[8,13]

Pathologic Effects

Pathologic effects of TMT are most prominent in the central nervous system (CNS) and produce neuronal necrosis, most prevalent in the hippocampus, amygdaloid nucleus, and pyriform cortex, rather than intramyelinic edema, which is specific for TET. Collectively, these areas are the "limbic system" that governs learning, memory, and behavioral responses to sensation.

Pathologic alterations in the mouse nervous system are dose-dependent; little pathology is observed at doses <3 mg/kg of body weight. However, a strong neuronal necrosis in the fascia dentata of the hippocampus appeared at 48 - 51 h after exposure to 3 mg/kg, when strain specific patterns of necrosis also appeared in the neocortex, pyriform cortex, amygdala nucleus, and brain stem. The large neurons in the brain stem showed most prominent damage.[12,15-17] Early post-natal mice show slightly different neuropathologic reactions.[18] In marmosets the neuropathology is similar.[19] In rats, as in mice, clearly defined areas of the brain were affected: the hippocampus, amygdala, and pyriform cortex.[8] However, Chang et al.[20] found species and strain differences among mice (BALB/c and C57BL/6) and rats (Longs-Evans and Sprague-Dawley) in TMT toxicity; brain regions sensitive to TMT toxicity also varied between the species.

Similar patterns in rat neuropathology after TMT administration were observed by Dyer et al.[13] and Ruppert et al.[21] Dense membrane-bound bodies appeared in close connection with rough endoplasmic reticulum (RER) and with the Golgi complex. It was suggested that the Golgi complex may be the seat of the critical metabolic lesion followed by disturbances in protein transfer and protein synthesis. For pigeons a TMT dose of 1.75 mg/kg of body weight produces almost total loss of hippocampal neurons, as well as neuronal damage in the brain stem and striatum griseum centrale.[22,23]

Pathologic changes in sensory nerves (retina, inner ear, pyriform cortex, olfactory tubercle, and dorsal root ganglia) as a result of TMT intoxication were investigated in Longs-Evans rats.[24] TMT retinopathy in rats was also recognized.[25]

Besides morphologic pathology, physiologic studies on neurotransmitter systems also demonstrated a number of significant TMT effects. In Longs-Evans rats brain uptake of the neurotransmitter GABA showed increased affinity and decreased capacity whereas uptake of norepinephrine showed reverse alterations.[26] In the same dose range of 3 mg/kg of body weight or less, the effects on dopamine and 3,4-dihydroxyphenylacetic acid (DOPAC) were marginal but at 7 mg/kg of body weight effects were decreased in the nucleus acumbens.[27] In post-natal rats GABA levels were reduced in the hippocampus and unchanged in the cortex and hypothalamus. Dopamine levels were reduced in the striatum but not affected in the brain stem. No change in DOPAC was observed in the striatum. No change was observed in norepinephrine in the brain stem or cerebellum, and acetylcholine was unaffected in the cortex, hippocampus and hypothalamus.[28]

Behavioral Effects

Behavioral alterations also characterize exposure to TMT. A marked whole-body tremor is reported for all species and strains. In Longs-Evans rats, TMT produces a "trimethyltin syndrome" characterized by spontaneous seizures, tail chasing, tail mutilation, and vocalization. Aggressive behavior was observed in Longs-Evans rats[13] and marmosets[19] but not in Sprague-Dawley rats,[29] C57BL/6N or BALB/c mice,[12] CFW or grasshopper mice[30] albino (TEX:ICR) mice,[31] or pigeons.[22,23,32]

TMT also affects motor activity. Among mice, intensity and duration vary by strain.[12]. For rats, increased activity was associated with increased water intake for 3 weeks following TMT administration[33] and with increased water intake, decreased food consumption, and altered diurnal rearing patterns.[34] In pigeons, motor activity was increased or decreased in a dose-dependent fashion.[35]

Several studies recognized effects of TMT on sensory function in laboratory animals. Fischer 344 rats demonstrated significant increases (25-60%) in latency to respond to a heat stimulus.[36] A five to sixfold latency was also reported in Longs-Evans rats.[37] Rats treated with TMT of 4, 5, 6, or 7 mg/kg of body weight showed dose-related changes in visually evoked responses that suggested retinal effects.[38] Young and Fechter[39] demonstrated auditory damage in rats after single injections of 2, 4, or 6 mg/kg of body weight TMT and a pattern of hippocampal damage similar to that caused by convulsant chemicals or seen in the brains of some human epileptics.[40,41]

Doctor et al.[42] studied effects of TMT on chemically induced seizures in mice. Following an intraperitoneal dose of 4.26 mg/kg of body weight and subsequent convulsant doses of bicuculine, INH, pentylenetetrazol, and strychnine, a decrease in seizure incidence for all compounds except strychnine was observed; this was followed at 16 h by spontaneous seizures and tremors followed by death. A dose-related decrease in seizure incidence and severity following electroshock was also observed after TMT administration.[43] In rats, induction of seizures had a latency of days.[44]

Other experimental studies tested the effects of TMT on conditioned behaviors. They showed that rats which received an oral 6 mg/kg of body weight dose required 1.5-2 times more trials than controls to complete a task.[45] Similarly, a significantly greater number of errors was found on a performance test in a Hebb-Williams maze.[46]

TMT also produced significant effects on operant behavior (lever pressing - food presentation) of Longs-Evans rats which persisted for long times.[47] Studies of conditioned behavior that required the subjects to choose between two or more stimuli decreased the matching accuracy of pigeons[32] and rhesus monkeys.[48]

Effects of Exposure During Development on Behavior

Behavioral studies also include exposures to TMT during prenatal and early neonatal periods. Ruppert et al.[49] gave TMT hydroxide at 0, 4, 5, or 6 mg/kg of body weight to rat pups on post-natal day 5. Rats given 5 or 6 mg/kg had a reduced body weight and an impaired rope-climbing ability at age 21 days. At about 120 days the animals showed a decreased startle response and a significantly different level of activity in a figure-8 maze. At 120 days of age, both the 5 and 6 mg/kg groups had decreased weights of the whole brain, olfactory bulb, and hippocampus.

In a developmental study female rats were given a chronic exposure to 0.15, 0.5 or 1.0 mg/liter of TMT in the drinking water from two weeks before breeding through post-natal day 21.[50] At age 11 days, only the pups from the 1.0 mg/liter dams showed a decrease in acquisition of the food-rewarded runway task. However, pups from all treatments showed a delay in extinction compared to pups from control dams.

Effects on People

Monitoring exposure of people to organotins is difficult since little information is available for judging the best indicators of exposure. In animal experiments, urinary and blood levels of tin are elevated with exposure. However, the rapidity of organotin clearance from blood in man may preclude the use of these measures in assessing toxicity for other than long-term exposure. Urinary tin and alkyl tin excretion studies may be more appropriate exposure measures for man.

Although the effects of organotin compounds on humans are relatively poorly known, several observations of poisoning or toxicity were reported. The first dramatic recognition of the toxicity of organotin compounds was in France in 1950 where inorganic tin compounds were used in a medicine for skin disorders. A commercial preparation of this medicine was contaminated with organotin compounds and resulted in more than 100 deaths.[51]

In 1951, Zeman et al. (cited by NIOSH[9]) reported four cases of exposure to an unknown concentration of TTMT and TET in a laboratory. The routes and duration of exposure could not be determined. The resulting symptoms included severe headache, nausea, and vomiting. The illness lasted 4-10 weeks. One case exhibited bradycardia, hypotension and arrhythmic heart beats, suggesting that either one or both of these two organotins are poisons of the circulatory system and autonomic nervous system.

Fortemps et al.[52] reported that two chemists, exposed to a mixture of dimethyltin and TMT from a closed system that leaked periodically, abruptly developed mental confusion, aggressiveness and seizures. The airborne concentrations of the alkyltins were never determined. Before the acute episode, the chemists had complained of headache, violent pain of the internal organs and psychic disturbances (memory loss, insomnia, anorexia, disorientation). According to the authors, the clinical evidence and exposure circumstances indicated poisoning by TMT. They attributed the acute incident to the cumulative neurotoxic effect of TMT. Both patients appeared to recover completely after removal from exposure.

More recently 22 workers were exposed to TMT over several months and given medical, neurological, psychiatric and neuropsychological exams.[53] While the exact amount of TMT exposure could not be determined, approximately one half of the group was present at the time and place of maximum exposure for a month. In this highly exposed group non-specific symptoms of central nervous system intoxication were increased significantly, and included forgetfulness, fatigue, sleep disturbances, and headaches. Unique and specific symptoms consisted of alternating attacks of rage and deep depression lasting a few hours to a few days. In 4 of the 12 individuals in the highly exposed group there were long-term personality changes. The exact physiologic mechanism producing these changes was not understood.

The unique pattern of limbic system toxicity of TMT was observed in all species examined to date. The involvement of the limbic system in behavior and memory strongly suggests that the behavioral effects of TMT are related to the pathology in this region of the brain. Based on the behavioral changes seen in human studies, it is likely that neuropathological changes were also present, although this was never determined. Unless extensive psychological testing and follow-ups are completed, this damage may go undetected, since individuals involved in TMT poisoning cases appeared outwardly normal and respond normally to physical and electrophysiological examinations (EEG). Sometimes only subtle tests of behavior are able to detect pathology associated with lesions in the limbic system of the brain. For example, rats and mice appeared normal after the acute behavioral symptoms of TMT toxicity subsided. However, examination of brain tissue revealed severe neuronal damage, and series of behavioral tests revealed serious deficits in memory and learning.

Because TMT binds to rat hemoglobin, it appears to be the species least sensitive (LD_{50}: 7-12 mg/kg of body weight) to the toxic effects of acute exposure to TMT. In contrast, since TMT does not bind to human hemoglobin,[14] estimate the lethal dose for humans at ~3.0 mg/kg of body weight or less. A limited number of reports on accidental exposures indicate that humans respond to a single acute dose or more alarmingly to repeated subtoxic doses of TTMT or TMT, suggesting a cumulative effect.

Diethyl Telluride

Toxicologic characteristics of diethyltelluride were evaluated by Kozik et al.[54] The LD_{50} dose for female and male rats, guinea pigs, and mice was determined as 55.0, 54.1, 45.1 and 154.0 mg/m^3, respectively. Although the values differed among the three species the compound was highly toxic by any measure. The compound had toxic effects on the blood, liver, kidney, heart, and skin and produced dystrophy in the small intestine. The effects of the compound were cumulative. Based on these studies, a maximum permissible concentration in the workplace air of 0.5 ug/m^3 was recommended.[54]

DISCUSSION AND CONCLUSIONS

Physical and biological data compiled in this report suggest that organometals must be used with caution if life and property are to be protected. Relative to the earth metals, the organic substitutes appear to present larger hazards. In this context, the actual incremental risks from their use will depend on a variety of factors including the basic integrity of process system as well as the operator interfacing needs. These will need to be more carefully evaluated in options such as metal organic chemical vapor deposition that will use large quantities of organometals and for which there is only limited operating experience. Special consideration should be given to handling systems and emergency response (e.g., fire-fighting) procedures. Risks to public health will probably decrease by substituting organometals for metal hydrides. Nevertheless, risk to occupational health and plant facilities will remain. As noted, use of organometals will require substantial revisions in engineering designs and operating procedures. Since there is only limited industrial experience available on which these systems are based, they should be scrutinized carefully and systematically. Although a casual review of the earth metals alone could indicate that the toxicity of these materials do not present problems to health, the available information for TMT and other organometals (e.g., methyl mercury and tributyltin) suggest that they should only be handled with great respect.

ACKNOWLEDGMENTS

This work was supported by the Photovoltaic Energy Technology Division, Conservation and Renewable Energy, under Contract NO. DE-AC02-76CHOOO16 with the U.S. Department of Energy. We thank A. Bulawka for his support and encouragement and A. Vanslyke for her help in typing and organizing this report.

REFERENCES

1. K. Rybicka, P.D. Moskowitz, L.D. Hamilton, Toxicology of tetramethyltin and other organometals used in photovoltaic cell manufacture. BNL Formal Report, Brookhaven National Laboratory, Upton, NY, 11973, (in press, 1988).

2. D.E. McMillan and G.R.Wenger, Pharmacol Rev. 37, 365 (1985).

3. D.C. Herron, Health hazards associated with tetramethyl tin. Internal ARCO Rep., Atlantic Richfield Co., Los angeles, CA, August 11, 1983.

4. R.H. Fish, E.H. Kimmel, and J.E. Casida, J. Organometal Chem. 118, 41 (1976).

5. E.C. Kimmel, R.H. Fish, and J.E. Casida J. Agric. Food Chem. 25, 1 (1977).

6. M.R. Krigman and A.P. Silverman, Neurotoxicology 5, 129 (1984).

7. S.V. Doctor, L.G. Sultatos, and S.D. Murphy, Toxicol. Lett. 17, 43 (1983).

8. A.W. Brown, W.N. Aldridge, B.W. Street, and R.D. Verschoyle, Am. J. Pathol. 97, 59 (1979).

9. NIOSH. Criteria for a recommended standard-occupational exposure to organotin compounds. U.S. Government Printing Office, Washington, D.C. (1976).

10. F. Caujolle, M. Lesbre, and D. Meynier, C.R. Acad. Sci. (Paris) 239, 556 (1954).

11. G.R. Wenger, D.E. McMillan, and L.W. Chang, Neurobehav. Toxicol. Teratol. 4, 157 (1982).

12. G.R. Wenger, D.E. McMillan, and L.W. Chang, Toxicol. Appl. Pharmacol. 73, 78 (1984).

13. R.S. Dye, T.J. Walsh, W.F. Wonderlin, and M. Bercegeay, Neurobehav. Toxicol. Teratol. 4, 127 (1982).

14. A.W. Brown, R.D. Verschoyle, B.W. Street, W.N. Aldridge, and H. Grindley, J. Appl. Toxicol. 4. 12 (1984).

15. G.R. Wenger, D.E. McMillan, and L.W. Chang, Toxicol. Appl. Pharmacol. 73, 89 (1984).

16. L.W. Chang, T.M. Tiemeyer, G.R. Wenger, and D.E. McMillan, Neurobehav. Toxicol. Teratol. 4, 149 (1982).

17. L.W. Chang, T.M. Tiemeyer, G.R. Wenger, D.E. McMillan, and K.R. Reuhl, Environ. Res. 29, 435 (1982).

18. K.R. Reuhl, E.A. Smallridge, L.W. Chang, and B.A. Mackenzie, Neurotoxicology 4, 19 (1983).

19. W.N. Aldridge, A.W. Brown, J.B. Brierley, R.D. Verschoyle, and B.W. Street, . Lancet 2, 692 (1981).

20. L.W. Chang, G.R. Wenger, D.E. McMillan, and R.S. Dyer, Neurobehav. Toxicol. Teratol. 5, 337 (1983).

21. P.H. Ruppert, T.J. Walsh, L.W. Reiter, and R.S. Dyer, Neurobehav. Toxicol. Teratol. 4, 135 (1982).

22. D.E. McMillan, M. Brocco, G.R. Wenger, and L.W. Chang, Toxicologist 2, 86 (1982).

23. D.E. McMillan, M. Brocco, G.R. Wenger, and L.W. Chang, Neurobehav. Toxicol. Teratol. 9, 67 (1987).

24. L.W. Chang and R.S. Dyer, Neurobehav. Toxicol. Teratol. 5, 676 (1983).

25. T.W. Bouldin, N.D. Goines, R.C. Bagnell, and M.R. Krigman, Am. J. Pathol. 104, 237 (1981).

26. J.J. Valdes, C.F. Mactutus, R.M. Santos-Anderson, R. Dawson Jr., and Z. Annau, Z., Neurobehav. Toxicol. Teratol. 5, 357 (1983).

27. D.L. DeHaven, T.J. Walsh, and R.B. Mailman, Toxicol. Appl. Pharmacol. 75, 182 (1984).

28. R.B. Mailman, M.R. Krigman, G.D. Frye, and I. Hanin, J. Neurochem. 40, 1423 (1983).

29. G.R. Wenger, D.E. Mcmillan, L.W. Chang, T. Zitaglio, and W.C. Hardwick Toxicol. Appl. Pharmacol. 78, 258 (1985).

30. K.L. Hulebak, and Z. Annau, Soc. Neurosci. Abstr. 7, 748 (1981).

31. L.G. Costa, S.V. Doctor, and S.D. Murphy, Life Sci. 31, 1093 (1982).41. K.L. Hulebak, and Z. Annau, Soc. Neurosci. Abstr. 7, 748 (1981).

32. S.O. Idemudia, D.E. McMillan, and L.W. Chang, Pharmacologist 25, 214 (1983).

33. C.T. Johnson, A. Dunn, C. Robinson, T.J. Walsh, and H.S. Swartzwelder, Neurosci. Lett. 47, 99 (1984.

34. P.J. Bushnell, and H.L. Evans, Toxicol. Appl. Pharmacol. 79, 134 (1985).

35. S.O. Idemudia and D.E. McMillan, Fed. Proc. 44, 494 (1985).

36. T.J. Walsh, R.L. McLamb, and H.A. Tilson, Toxicol. Appl. Pharmacol. 73, 295 (1984).

37. W.E. Howell, T.J. Walsh, and R.S. Dyer, Neurobehav. Toxicol. Teratol. 4, 197 (1982).

38. R.S. Dyer, W.E. Howell, and W.F. Wonderlin, Neurobehav. Toxicol. Teratol. 4, 191 (1982).

39. J.S.Young and L.D. Fechter, Toxicol. Appl. Pharmacol. 82, 87 (1986).

40. R.S. Dyer, Neurobehav. Toxicol. Teratol. 4, 659 (1982).

41. R.S. Dyer, W.F. Wonderlin, T.J. and Walsh, Neurobehav. Toxicol. Teratol. 4, 203 (1982).

42. S.V. Doctor, L.G. Costa, and S.D. Murphy, Toxicol. Lett. 13, 217 (1982).

43. S.V. Doctor and D.A. Fox, J. Toxicol. Environ. Health 10, 43 (1982).

44. R.S. Sloviter, C. von Knebel Doeberitz, T.J Walsh, and D.W. Dempster, Brain Res. 367, 169 (1986).

45. T.J. Walsh, D.B. Miller, and R.S. Dyer. Toxicol. Teratol. 4, 177 (1982).

46. H.S. Swartzwelder, J. Hepler, W. Holahan, S.E. King, H.A. Leverenz, P.A. Miller, and R.D. Myers, Neurobehav. Toxicol. Teratol. 4, 169 (1982).

47. H.S. Swartzwelder, R.S. Dyer, W Holahan, and R.D. Myers, Neurotoxicology 2, 589 (1981).

48. P.J. Bushnell and H.L. Evans, Fed. Proc. 43, 763 (1984).

49. P.H. Ruppert, K.R. Dean, and L.W. Reiter, Neurobehav. Toxicol. Teratol. 5, 421 (1983).

50. E.A. Noland, D.H. Taylor, and R.J. Bull, Neurobehav. Toxicol. Teratol. 4, 539 (1982).

51. J.M. Barnes, and H.B. Stoner, Pharmacol. Rev. 11, 211 (1959).

52. E. Fortemps, G. Amand, A. Bomboir, R., Lauwerys, and E.C. Laterre, Int. Arch. Occup. Environ. Health 41, 1 (1978).

53. W.D. Ross, E.A. Emmett, J. Steiner, and R. Tureen, Am. J. Psychiarty 138, 1092 (1981).

54. I.V. Kozik, N.P. Novikova, L.A. Sedova, and E.N. Stepanova, (Russian) Gig. Tr. Prof. Zabol. 10, 51 (1981).

Table I. Earth metals, metal binary hydrides and organic substitutes used in photovoltaic cell manufacture.

Earth metal Metal binary hydride	Earth metal/ Organic substitute	CAS Registry No.
Aluminum	Trimethylaluminum	00075-24-1
Gallium	Trimethylgallium	01445-79-0
Hydrogen telluride	Diethyltelluride	00627-54-3
Indium	Trimethylindium	03385-78-2
Tin	Tetramethyltin	00594-27-4
Zinc	Diethyl zinc	80557-20-0
	Dimethyl zinc	00544-97-8

Table II. Physical and chemical properties of organometals used in the photovoltaic industry.

Compound	Formula	MW	Point (C^o) Melting	Point (C^o) Boiling	Appearance and odor	Vapor pressure (mmHg)
Diethyl-telluride	$(C_2H_5)_2Te$	123	unknown	138	Liquid garlic odor	
Diethylzinc	$(C_2H_5)_2Zn$	124	-28	117.6 118	Liquid colorless	15 @ 20°
Dimethylzinc	$(CH_3)_2Zn$	95	-46	46	Liquid colorless	123 @ 0°
Trimethyla-luminum	$(CH_3)_3Al$	72	0	130	Liquid colorless	9.2 @ 20°
Trimethyl-gallium	$(CH_3)_3Ga$	115	-19	55.7	Liquid colorless	64.5 @ 0°
Trimethyl-indium	$(CH_3)_3IN$	160	90		Crystals white	
Tetrame thyltin	$(CH_3)_4Sn$	179	-54	78	Liquid colorless musty odor	

Table III. Hazardous properties of organometals used in the photovoltaic industry.

Chemical	Fire and explosion hazard
Diethyltelluride	Self ignites in air
Dimethylzinc	Spontaneously flammable in air; violent reactions with air, water, Cl_2 hydrazine
Trimethylaluminum	Spontaneous reaction with air explodes with water
Trimethylgallium	Spontaneously flammable in air
Trimethylindium	Flammable and pyrophoric; may explode in contact with water
Tetramethyltin	Highly flammable (no spontaneous ignition) unreactive with water

MEDICAL SURVEILLANCE AND BIOLOGIC MONITORING OF PERSONNEL IN SEMICONDUCTOR RESEARCH FACILITIES

G. Krieger, MD, MPH and R. Cambridge, RN
Solar Energy Research Institute
Golden, Colo. 80401

ABSTRACT

Semiconductor research facilities have many of the same health and safety hazards as large-scale manufacturing plants. The novel and episodic nature of laboratory activity requires a different perspective for medical surveillance and biologic monitoring. Basic principles of medical surveillance for laboratory personnel are reviewed and analyzed. Approaches to biologic monitoring and the role of biologic exposure indices are discussed. Specific types of hazards including dopants, gallium arsenide, toxic inhalations, and physical hazards are placed in a framework consistent with laboratory exposures. Reproductive effects and carcinogenic potential of selected substances are assessed along with possible monitoring strategies.

INTRODUCTION

Laboratory facilities are often viewed as safe and intellectual environments that do not require high degrees of health and safety surveillance. This attitude is widespread despite a) the large number of chemical and physical hazards, b) the high toxicity potential of new compounds and processes, c) the rapid change and introduction of new equipment, and d) the use of exotic materials that have been minimally assessed for human health impact. To compound this overall problem, semiconductor production facilities, even at large-scale manufacturing sites, have a largely undeserved reputation as super clean and rigorously controlled environments. This perspective has been reinforced by several studies of semiconductor manufacturing sites.[1-5] The prevailing attitude has been that the actual manufacturing processes require stringent control over possible accidents and this results in a high degree of environmental containment that automatically reduces possible worker exposure. A recent review of occupational injury and illness statistics in the California semiconductor industry belies the optimistic view of previous studies.[6] Furthermore the large-scale control technology used in plant operations does not necessarily have laboratory size equivalents. Personnel in small or individual research laboratories can be particularly vulnerable to leaks, spills, and exposures generated during sample preparation or equipment maintenance and cleaning. Thus, medical surveillance and biologic monitoring techniques that are applicable to a large-scale manufacturing facility may be irrelevant to the episodic starts and stops of laboratory research.

BASIC PRINCIPLE OF MEDICAL SURVEILLANCE IN THE LABORATORY SETTING

Laboratories tend to operate in a nonlinear fashion; specifically, they generally do not run continual nonchanging processes. The starting materials and physical hazards such as infrared (IR), ultraviolet (UV), and laser use can and will vary from experiment to experiment. Furthermore, the time-table of experiments can fluctuate widely and include nights, holidays, and weekends. Thus, the timing of lab visits must match the time schedule of activity for that particular department. Otherwise standard industrial hygiene or medical "walk throughs" may provide little useful information.

A similar problem can occur with standardized chemical inventories. We use these listings and they can be helpful, but are prone to problems of incompleteness and nonreporting of new substances in a particular laboratory's inventory. As research teams pursue promising levels or discover blind-ends the type and quantity of hazardous materials can fluctuate dramatically.

We also review each laboratory's safe operating procedures or "SOPs." These are written specifically for the experiment by the research team during the design of the process. These are helpful in understanding the thrust of the experiment and in familiarizing us with the hazardous aspects. As with the chemical inventory, problems arise when the process is modified and the SOP is not updated.

Currently, it is our practice to perform a scheduled site visit at each lab and with each employee on an annual or semi-annual schedule. Lab visits are scheduled during regular operations and particularly in the time frame of new experiments or processes. To facilitate this process, we have developed a standardized form to use in data collection (see Table I). Additionally, lab personnel are interviewed in two different environments: a) at their lab during the site visit, and b) in the medical department in a confidential setting. After the site visit a preliminary plan for medical surveillance is formulated. During the final interview with the employee, the plan is reviewed, discussed, and implemented. This process allows full employee participation, confidentiality and individualization. Thus, the best historical data is obtained and the timing of biologic testing fits real-time exposure parameters rather than keying examination to irrelevant milestones such as birth or employment dates.

PRINCIPLES OF BIOLOGIC MONITORING IN A LABORATORY SETTING

Biologic monitoring is the process of assessing chemical exposure by measuring specific substances in biologic specimens. These specimens are generally blood, urine, and exhaled air. Other specimens include: feces, hair, nails, breast milk, saliva, semen, or biopsied fat. Each of these specimens has particular advantages in certain monitoring situations; however, problems in sample collection storage and analysis can be significant. Because of the wide variety and unusual nature of many of the substances found in the semiconductor industry, the monitored substances should meet certain broad criteria: 1) the specimen should be easily obtainable, such as blood or urine, 2) analytic methodology should exist to measure the specific substance in the given specimen accurately, 3) the results should be interpreted in terms of published reference limits.

Urine is the specimen that generally satisfies these conditions in most routine circumstances. Urine contains nearly all the xenobiotics in amounts generally proportional to the absorbed dose. It is our practice to obtain 24-hour specimens rather than spot or 2-hour samples. In a large-scale plant operation 24-hour urines are impractical but for small laboratory exposures they are ideal.[7,8]

TABLE I LABORATORY EVALUATION FORM

EMPLOYEE ______________________ LOCATION ______________________

HAZARDS:

RADIATION ____________________ SPECIALTY GASES ____________________

KINETIC ENERGY ________________ CARCINOGENS ____________________

ELECTRICAL ____________________ TOXIC SUBSTANCES ________________

HIGH/LOW TEMP _________________ CORROSIVE AGENTS ________________

BIOLOGICAL ____________________ HIGH/LOW PRESSURE _______________

NOISE _________________________

SOURCES RELATIVE TO ABOVE:

______________________________ ______________________________

______________________________ ______________________________

______________________________ ______________________________

RECOMMENDED CONTROL STANDARDS:

____

____

INDUSTRIAL HYGIENE REFERRAL: YES ____________ NO ____________

BACKGROUND INFORMATION:

EXPOSURE POTENTIAL (LOCATION & TIME) ____________________________

____

PERSONAL PROTECTIVE EQUIPMENT ____________________________________

____

ENGINEERING CONTROLS ____

____

MEDICAL SCREENING ____

____

AREA/PERSONAL MONITORING _______________________________________

____

TRAINING REQUIREMENTS ____

____

SOP REVIEW ____________________ CHEM INV REVIEW ________________

MEDICAL DEPARTMENT SIGNATURE EMPLOYEE SIGNATURE

______________________________ ______________________________

DATE __________________________ DATE __________________________

This contrasts with the problems associated with blood. While blood levels can show high correlation with ambient environmental concentrations there are three major problems: 1) it is difficult to obtain frequent specimens, 2) low exposure-levels produce blood concentrations below analytic detection limits, and 3) blood is a difficult analytic matrix for analysis.

Our current practice is to use 24-hour urine collections whenever possible and to reserve blood specimens for yearly general chemistry and hematologic surveillance.

Biologic exposure indices (BEIs) are levels of determinants observed in specimens collected from an otherwise healthy worker who has been exposed to chemicals. This is similar to the monitoring of a worker with an inhalation exposure to the standard 8-hour time weighted average (TWA or also TLV [threshold limit value]).[9] BEIs provide a method to directly compare exposure levels in humans similar to industrial hygiene monitoring. BEIs have been established for only a limited number of substances and this reduces their usefulness. However, current research directions should increase their availability and utility. Given the episodic nature of laboratory exposures, BEIs are useful because they correlate exposure intensity with the absorption, elimination, and metabolism of chemicals in humans.

Our overall approach is to view laboratory monitoring as a pyramidal structure (see Table II). At the base of the pyramid is the review of chemical inventories and safe operating procedures. Next are the general blood chemistry and hematologic tests. Our next level consists of more specific tests for direct exposure. These include 24-hour urine levels for specific substances such as arsenic or selenium, or BEIs for solvents such as toluene, xylene, or trichlorethylene. Our final level of monitoring involves tests for mutagenicity, carcinogenicity, and altered immune function.[10] These tests include sister chromatid exchanges (SCEs), T-cell function studies, and Ames tests on concentrated urine. In a workplace, these tests are clearly controversial; however, they do allow a sensitive assessment of cellular functions not measured by traditional testing schemes. We anticipate that these sophisticated tests will become increasingly common in the workplace through the next 5-10 years.

SPECIFIC CATEGORIES OF HAZARDS

Any process that exists at a large-scale manufacturing facility had its genesis as a laboratory experiment. Hence the range of possible chemical and physical hazards is tremendous. In the last three years, there have been excellent review articles with extensive lists of chemicals, gases, physical hazards, and manufacturing processes.[11-17]

In general, seven specific categories of hazards should be evaluated (see Table III). Appropriate biologic monitoring techniques are generally available for each category (see Table II). Certain types of problems are fairly common in photovoltaic laboratories and should be specifically questioned during interviews with lab personnel (see Table IV). In our experience, the majority of exposures occur during routine maintenance and cleanup rather than during the actual experiment. Therefore, biologic monitoring should be timed to correspond to these activities as potentially significant exposures can occur.

Table II STRUCTURE OF MEDICAL SURVEILLANCE MONITORING

1. Review of chemical inventory and safe operating procedures

2. General blood chemistry, hematology analysis, chest X-rays, pulmonary function tests, urinalysis, and baseline physical examination and history

3. Specific chemical analysis, BEIs, site visits, and site surveys

4. Tests for carcinogenicity, immune function, and mutagenicity

TABLE III SPECIFIC HAZARDS

1. Dopants and their related compounds.
   Antimony, arsenic, boron, cadmium, indium, phosphorus, selenium, tellenium, zinc

2. Gallium Arsenide and its related process technology

3. Gases with Toxic Inhalation Effects
   Ammonia, arsine, diborane, hydrogen chloride, phosphine, silane

4. Physical Hazards
   Microwave, laser, radio frequency, infrared, ultraviolet, X-ray, electron beam

5. Solvents
   Chlorinated hydrocarbons, alcohols, nonchlorinated hydrocarbons

6. Carcinogenic Substances
   Solvents - trichloroethylene
   Metals - arsenic, cadmium, chromium

7. Reproductive Hazards
   Metals - arsenic, cadmium
   Solvents - 1,1,1-trichloroethane
   Photoresists - methoxyethanol, ethoxyethanol
   Physical Agent - ionizing radiation

TABLE IV PROBLEM AREAS IN PHOTOVOLTAIC LABORATORIES

1. Open reactors between runs
2. Gas tank changes
3. Reactor changing
4. Laser and RF exposure
5. Hood performance and utilization
6. Lab ventilation and exhaust system
7. Wet etching
8. Any use of hydrofluoric acid
9. Mineral acids and caustics
10. Epoxy resins
11. Vapor leaks with contamination into vacuum pump oils
12. Solvent exposure during cleanup, maintenance, and sample preparation.

CONCLUSION

Photovoltaic laboratories and research facilities have the potential for significant exposures to a wide variety of chemical and physical hazards. Medical surveillance and specific biologic monitoring must be layered and timed to accurately reflect the noncontinuous nature of the laboratory environment. This process can only be accomplished by a proactive approach that includes frequent site visits and constructively involves the laboratory personnel.

REFERENCES

1. NIOSH (1977): Health Hazard Evaluation Determination No. 77-01-376, FMC Co., Broomfield, CO. Cincinnati: DHEW.
2. NIOSH (1980): Health Hazard Evaluation Determination No. 79-66. Signetics Co., Sunnyvale, CA. Cincinnati: US DHHS.
3. NIOSH (1980a): "Industrial Hygiene Characterization of the Solar Cell Industry." Technical Report 80-112 Cincinnati: US DHEW.
4. NIOSH (1985): "Hazard assessment of the electronic component manufacturing industry." Report No. 85-100. Cincinnati: US DHHS.
5. R. Wade (1981) "Semiconductors Industry Study." California Department of Industrial Relations, Task Force on the Electronics Industry.
6. J. Ladou. State of the Art Reviews: Occupational Medicine. 1(1):1 1986.
7. W. J. Turner. Clin Pharm Ther. 12:163, 1971.
8. R. Louwerys. Scand. J. Work Env. Health 1:139, 1975.
9. ACGIH: "Threshold limit values and biologic exposure indices for 1987-88" Cincinnati: American Conference of Governmental Industrial Hygienists.
10. G. M. Williams. Casarett and Doull's Toxicology (Macmillan, N.Y., 1986), p. 99.
11. P. H. Wald. Am. J. Ind. Medicine 11:203. 1987.
12. J. LaDou (ed.) Microelectronics Industry. State of the Art Reviews (Hamley and Belfus, Philadelphia) 1:1. 1986.
13. V. M. Fthenakis. Solar Cells, 13:43. 1984.
14. J. C. Lee. Rep. BNL 51768 (Oct. 1983)
15. V. M. Fthenakis BNL 51854, Sept. 1984.
16. V. M. Fthenakis BNL 51853 April 1985.
17. P. D. Moskowitz. BNL 51832 April 1985.

THE ROLE OF UNUSUAL OCCURRENCE REPORTS IN INCIDENT ANALYSIS IN THE PHOTOVOLTAIC INDUSTRY

S. Thompson
Solar Energy Research Institute, 1617 Cole Blvd., Golden, Colo. 80401

ABSTRACT

Incidents affecting operational performance within the photovoltaic industry can be reduced by implementing unusual occurrence reporting systems. This paper identifies the major components of such a system, including background and criteria for determining an unusual occurrence, components of an unusual occurrence report form, and the process cycle from an occurrence to its resolution. Examples of occurrences at the Solar Energy Research Institute (SERI) are presented along with their final results.

INTRODUCTION

There are many hazardous technologies employed by photovoltaic industries that can severely affect programmatic operations when a major unplanned event occurs. An unusual occurrence is an unplanned event that significantly affects the programmatic operation, safety, or reliability of a facility. An unusual occurrence reporting system is employed by SERI and other Department of Energy (DOE) contractors as a management tool to increase safety awareness among staff members. The reporting system expands management's understanding of significant operational issues and problems.

To be classified as an unusual occurrence, an event must be significant. Individual groups should define their own criteria, but should include categories addressing unanticipated costs; program delays; reduced efficiency, performance, and reliability; and unsafe activities. The primary component of this type of reporting system is the unusual occurrence report (UOR). The UOR is generated by the personnel who are most responsible for a problem area, such as researchers in a laboratory setting. The report is then reviewed and evaluated by senior management to determine the consequences of the occurrence and its impact on operations. In addition, all parties are responsible for determining the root cause of the incident and for proposing creative solutions to the problem. Root causes often are not identified because some groups emphasize assessing blame rather than determining the underlying causes of the occurrence. Therefore, many people will withhold the necessary information, fearing disciplinary action or embarrassment. The UOR concept deemphasizes blame by attempting to identify the cause of an occurrence, correct deficient areas, and share the final results with affected parties. The information sharing process has been particularly beneficial because the lessons learned have allowed organizations to avoid similar problems within their own operations.

TYPES OF UNUSUAL OCCURRENCES

Incidents that can be classified as unusual occurrences are not confined to those involving damage to personnel or property. Following are examples of typical incidents that could adversely affect the programmatic goals and objectives of a business.

- Deviation from approved safe operating requirements
  - Unauthorized use of unsafe or dangerous processes
  - Failure of personnel to use safety equipment
  - Bypassing of safety systems and interlocks
  - Operation of equipment or processes above maximum allowable limits
- Failure of a protective system
  - Failure of safety alarms or signals, including facility alarms
  - Unauthorized or unknown shutdown of sprinkler systems
  - Inadequate performance of emergency power systems during an emergency
- Malfunction of a facility, operation, or system component
  - Unanticipated rupture leakage or equipment degradation
  - Malfunction of fire protection systems
  - Failure of beams, concrete, rags, or similar items
- Insufficient specifications
  - Inadequate fire protection equipment
  - Insufficient personnel barriers or machine guards
  - Use of improper materials
  - Failure of prototype equipment or systems during testing
  - Inaccurate or inadequate information pertaining to design requirements
  - Improper design, manufacture, or performance that endangers a facility or major program
- Natural or man-made events
  - Unstable soil conditions that threaten the structural integrity of facilities or major utilities
  - Interruption of a water supply to a sprinkler system
  - Penetrations of fire walls that significantly increase risk to personnel or facilities
  - Inclement weather that damages facilities
  - Restricted emergency access or egress from facilities because of man-made or natural barriers
  - Human errors that could threaten the performance, reliability, or safety of operations
- Fire or explosions.

THE UOR FORM

UOR forms must contain standard information so that personnel can investigate the incident adequately. SERI generates reports that contain specific information such as the UOR number, location of the occurrence, status and date of report, division or project, operating conditions at the time of the incident, initial observations made, immediate corrective action taken, report status, final evaluation, lessons learned, and programmatic impact. The items that are not self-explanatory are addressed in the following sections.

THE UOR NUMBER

Organizations must develop internal tracking of UORs. This tracking should be the responsibility of a designated group, preferably the company's safety and health department, in order to ensure program consistency. The assigned group will issue the appropriate UOR number, monitor the status of the report, and obtain proper closure.

OBSERVATIONS MADE AT TIME OF INCIDENT

It is important to record any unusual conditions or activities immediately. Experience has shown that details initially judged to be insignificant may eventually prove valuable to the investigation.

FINAL EVALUATION AND LESSONS LEARNED

The final disposition of the occurrence should be documented and the lessons learned communicated to affected parties. The sharing of lessons learned is the key element in an effective UOR system. The lessons learned can help prevent a similar incident in the future. The findings of the UOR should be shared with individuals who perform similar functions, thereby increasing safety awareness and enhancing understanding of operational concerns. DOE requires that unusual occurrences are reported to the appropriate operations office. The reports are then placed in a central computer for distribution to appropriate DOE facilities. This process benefits personnel at different work sites who may have limited contact with their peers in similar industries.

MANAGEMENT REVIEW

All UORs must be reviewed and signed by the originating division. Reviewers must ensure that each unusual occurrence is described clearly and concisely. Often, drawings and charts can clarify the description. In addition, the report should state the cause, immediate action taken to reduce the negative consequences, and the short- and long-term effects on the operation of the facility. The reviewer should also determine that the recommended corrective actions adequately address the root causes in order to prevent recurrence of the incident.

After the division's review, the UOR should be returned immediately to the monitoring group for internal and external distribution, if applicable. Affected organizations should also submit the document to internal safety committees to increase awareness.

THE UOR PROCESS

SERI has incorporated specific procedures to facilitate the UOR process by ensuring consistency among the many potentially affected groups. These procedures not only ensure internal programmatic consistency, but provide a means of complying with DOE requirements.

The first step in the UOR process is the immediate notification of the cognizant management personnel. Management should be notified as soon as the nature and extent of the occurrence are known. DOE requires that unusual occurrences are reported to area offices immediately.

Once the proper notifications have been made, an evaluation of the situation must begin. The goal of the investigation is to identify the cause of the occurrence, its impact on organizational programs, and the appropriate corrective action. Investigations should be probing, but should avoid placing blame on individuals. Placement of blame can be avoided by addressing areas such as inadequate personnel training, design modifications, or revision of operating procedures. Investigators should concentrate on hidden causes instead of becoming preoccupied with obvious causes. For example, if an employee was working in a laboratory and splashed a chemical in his or her eye, the cause of the occurrence would probably be identified as the individual's failure to wear proper eye protection. However, the hidden cause might be an inadequate management system. The management system might not have identified the need for personal protective equipment such as goggles, a face shield, gloves, or aprons. Another possibility is that management did recognize the necessity of such equipment but did not ensure that it was issued and properly used by employees. Only by identifying these types of hidden causes can investigations be justified as essential to preventing a recurrence.

The final step in the UOR process is submitting a written report. All identified unusual occurrences at SERI must be noted in writing in order to keep a record of the occurrences. Reports must be written so that reviewers unfamiliar with operations, equipment, and facilities can comprehend and analyze the situation. Any proprietary information or defense-related data should be deleted from the report.

Written UORs can fall into three categories: initial, interim, and final. Initial reports are usually submitted within 10 working days. Interim reports may need to be generated to address the status and progress of the investigation. Final reports are issued once corrective action has been completed by the originating group or division. The final document should contain pertinent information from previous reports to provide a complete overview of the occurrence. It should also include an evaluation of the hidden cause(s), completed corrective action, and ongoing activities implemented to prevent recurrence.

EXAMPLES OF UNUSUAL OCCURRENCES

The following items are actual unusual occurrences that have taken place at SERI.

SUPPLY AIR FAN FAILURE

A Joy Manufacturing Company Series 1000, Model 1770 supply air fan was operating in the east fan loft of SERI's Field Test Laboratory Building (FTLB) when it suddenly disintegrated. At the time of the incident the building's operating conditions were normal. Maintenance personnel had reported an unusual noise coming from the fan earlier in the day. Joy Manufacturing was contacted about the noise, but advised SERI that the sound was standard.

The maintenance personnel inspected the fan immediately after it was destroyed and determined that it did not pose a threat to personnel or property. The immediate evaluation stated that the fan probably suffered from inadequate bearing design and/or balance. Examinations of similar units revealed the need to replace the bearings in two fans and led to the installation of vibration sensors and alarms on all fans. Further investigation revealed that Joy Manufacturing Company had stopped using the type of bearing that was in SERI's unit. The company had also revised its recommended lubrication procedure because Joy considered the one that SERI had been following to be "over greasing." Unfortunately, Joy had not communicated the revision to SERI.

The final evaluation revealed that a factory defect caused the destruction of the fans. SERI personnel also modified internal procedures so that improperly functioning units are immediately removed from service and evaluated. SERI provided the lessons learned to other DOE facilities, some of which used the same type of unit. These facilities made the proper adjustments, which prevented additional losses.

EXPLOSION IN MULTI CHAMBER RESEARCH SYSTEM

An explosion occurred in the multi chamber amorphous silicon deposition research system, which consists of experimental equipment being used to fabricate thin-film amorphous semiconductors. The preliminary description of the occurrence revealed that as silane was being introduced into one of the chambers, the silane either reacted with an air leak or was introduced at an unusually high flow rate that caused an explosion within the chamber. Further investigations showed that the initial quantity of silane passed too quickly through the effluent gas scrubber, causing the silane to react with the oxygen at the exhaust end and burn within the exhaust piping. The piping, made of polyvinyl chloride (PVC), partially melted. The burning was brought under control as the silane concentration in the nitrogen ballast was reduced.

The deposition system operators immediately noted that the safety interlocks closed all valves leading in and out of the chamber, thereby isolating the silane gas leading from the cylinder to the chamber. The MDA toxic gas alarm was then monitored to determine if a toxic leak had developed. Subsequently, the affected chamber was pumped down under high vacuum, securing the operation from further damage.

Unfortunately, the exact cause of the occurrence may never be learned. However, the PVC piping was replaced with steel pipe. If the PVC piping had completely melted, it would have exposed the fire to the laboratory, causing a severe danger of a more widespread fire. Replacing the PVC with steel pipe should lessen this fire danger.

One possible cause could be that the chambers were constructed of aluminum instead of stainless steel, which is the industry standard. However, additional research is needed before a proper analysis can be made.

FACILITY FIRE

The contents of a drum of miscellaneous chemicals were ignited by a contractor who was using a cutting torch to dismantle a building. The contents of the drum were not immediately hazardous because of the fire, but there were boxes nearby that contained more than 40 different types of chemicals including 2 gallons of isopropyl alcohol, 2 gallons of hydrogen peroxide, and 3 pounds of powdered magnesium. SERI personnel removed the drum from the building with a forklift before the fire spread to the boxes.

The cause of the accident was determined to be a failure to follow established safety procedures on two counts. First, the contractor was not issued a cutting and welding permit by SERI Safety and Health personnel. The permit would have specified that any flammable materials nearby had to be moved. Second, the storage of combustible chemicals violated safe operating practices and constituted an unusual occurrence in itself. The situation would have been more serious had the boxed chemicals come into contact with the fire. Because of this incident, SERI revised its policy dealing with contractors to include the requirement for a prework safety orientation.

CONCLUSION

Implementing a UOR system can benefit all industries and is not only confined to safety issues. Organizations should determine their own specific requirements and tailor the system to their individual needs.

The difference between UORs and traditional accident investigations is that UORs are generated by the personnel most directly involved with the occurrence. Their input is invaluable in determining the root causes of unusual incidents. The cooperation of the involved individuals can be obtained by emphasizing that the UOR system will not be a fault-finding mission. Investigators should concentrate on programmatic inadequacies and recommend proper corrective action.

UOR systems have reduced occurrences at many facilities because lessons learned were shared with others, allowing modifications to prevent a similar occurrence. Many companies could address their own internal communication problems by initiating a UOR system.

BIBLIOGRAPHY

U.S. Department of Energy, 1981, *Unusual Occurrence Reports*, Order DOE 5000.3, Washington, D.C.: U.S. Government Printing Office.

U.S. Department of Energy, 1981, *Unusual Occurrence Report System*, Order DOE 5484.2, Washington, D.C.: U.S. Government Printing Office.

CONTROL OF ACCIDENTAL RELEASES OF HYDROGEN SELENIDE IN VENTED STORAGE CABINETS

V.M. Fthenakis and P.D. Moskowitz
Brookhaven National Laboratory, Upton, New York 11973

R.D. Sproull
Oregon State University, Corvallis, OR 97331-2702

ABSTRACT

Highly toxic hydrogen selenide and hydrogen sulfide gases are used in the production of copper-indium-diselenide photovoltaic cells by reactive sputtering. In the event of an accident, these gases may be released to the atmosphere and pose hazards to public and occupational safety and health. This paper outlines an approach for designing systems for the control of these releases given the uncertainty in release conditions and lack of data on the chemical systems involved. Accidental releases of these gases in storage cabinets can be controlled by either a venturi and packed-bed scrubber and carbon adsorption bed, or containment scrubbing equipment followed by carbon adsorption. These systems can effectively reduce toxic gas emissions to levels needed to protect public health. The costs of these controls (~$0.012/W_p$) are small in comparison with current (~$6/W_p$) and projected (~$1/W_p$) production costs.

INTRODUCTION

Accidental releases of toxic gases and options to mitigate them are receiving great attention after recent tragedies involving toxic gas releases. For many of the gases used by the industry there are not established prescribed control options. This is especially true for hydrogen selenide which is used in the manufacturing of copper-indium-diselenide photovoltaic cells by reactive sputtering. The gas has not been used in large quantities in the past and data on its chemical interactions with potential neutralizers are lacking. The objective of this paper is to specify environmental control systems for facilities using this gas and evaluate alternatives, given the uncertainty of many of the design parameters. Since hydrogen selenide will coexist with hydrogen sulfide the latter has also been considered in specifying these control systems.

RELEASE SCENARIOS

Material flows in a hypothetical 10 MW_p per year manufacturing facility were used as a basis of this study. The annual H_2Se and H_2S requirements for this plant have been estimated as 1590 kg and 650 kg, respectively.[1] We assume that the quantities of these gases stored on site would correspond to a single week's supply, i.e., 32 kg of H_2Se and 13 kg of H_2S (2 cylinders from each gas). In this study control systems which have the capacity to control the total release of the on-line feedstock (e.g. one full cylinder) are specified. We have considered five different release scenarios: (i) ruptured cylinder,

(ii) punctured cylinder, (iii) melting of the fusible plug, (iv) regulator failure, and (v) leak in lines to and from operating equipment.

The worst scenario involves the release of the contents of a full cylinder, for example the rupture of a single cylinder with the instantaneous escape of 16 kg of H_2Se or 6.5 kg of H_2S. If this occurs inside a gas storage cabinet, the sudden expansion of gas would probably cause gas to leak out of the cabinet and into the workspace. Although this scenario cannot be easily controlled, it is also the least likely to occur. A more realistic accident involves the puncturing of a cylinder or melting of its fusible plug and leakage of its contents in a period of 30 s to 30 min. This represents the "worst credible" scenario assumed in our analysis. This release then must be contained and diverted into a control system designed to handle such releases. The other scenarios involve smaller flow rates of release which can be handled by the same system. Using flow-restricting values in the cylinders substantially reduces the leakage flow through a cylinder valve (e.g. using a flow restricting valve with an orifice of 6 mills in a gas cylinder at 2000 psig pressure will reduce the flow out of the cylinder from about 2500 lpm to 30 lpm). A distribution line leak equipment could release a maximum of 3.7 g/min of H_2Se and 1.5 g/min of H_2S. Similarly, leaks in the lines off the reactors could release 2.2 g/min of H_2Se and 0.9 g/min of H_2S. These line leaks can be controlled by automatic shut-off valves activated by gas detectors and purging of released gas to the control equipment through race ways enclosing the pipes.

EMISSION STANDARDS

There are no federal emission standards for the routine or accidental release of these toxic gases from manufacturing facilities. If this facility were to be constructed in New York State, however, emission regulations for toxic gases promulgated by the New York State Department of Environmental Conservation (NYSDEC) would apply.[2] Accidental releases lasting more than one hour are considered routine emissions and the allowable limits for routine emissions apply. For releases of shorter duration, the NYSDEC specifies that the 15-min average emission concentrations must be less than the Treshold Limit Values (TLVs), as specified by the American Conference of Governmental Industrial Hygienists.

For routine emissions NYSDEC guidelines specify an Acceptable Ambient Level (AAL) of 0.01 ppm for H_2S, and 0.0002 ppm for H_2Se. A factor of 100 is allowed for dispersion of the gases from the stack to the receptor, and therefore, the in stack allowable limits during routine operation for H_2Se and H_2S are 0.02 ppm and 1 ppm, respectively (see Table I).

POLLUTION CONTROL OPTIONS

Several environmental control options exist for the treatment of H_2Se and H_2S including wet scrubbing, carbon adsorption, and combustion. The selection of the system depends on the type and quantities of gases to be controlled and the required level of control. Both scrubbing and adsorption are viable alternatives, but the toxicity of H_2Se and H_2S combustion products (mostly SeO_2 and SO_2) eliminates combustion from further consideration. In addition it is more difficult to scrub the oxidation products than the original gases.

Wet Scrubbing

A solution typically used for scrubbing H_2S is 1 M sodium hydroxide. It forms sodium sulfide according to the basic reactions:

$$H_2S(g) + NaOH(aq) \longrightarrow NaHS(aq) + H_2O$$

$$NaHS(aq) + NaOH(aq) \longrightarrow Na_2S(aq) + H_2O$$

These reactions occur rapidly in the aqueous phase under standard conditions. The reactions of H_2Se are expected to be similar. The equilibrium constants for the reactions of H_2S and H_2Se with NaOH solutions are 1.2×10^5 and 1.8×10^{15}, respectively. Because kinetics, transport, and solubility data for H_2Se are lacking, H_2S properties are used to approximate H_2Se values. The diffusion coefficient of H_2Se in air is estimated to be about 75% that of H_2S.[3] The overall rate of reaction of $H_2S(g)$ with 1 M NaOH(aq) depends on both the absorption rate of the gas into the liquid and the rate of reaction. However, since reactions occur rapidly under standard conditions, the gas-phase mass-transfer coefficients can be used in a preliminary design, as a surrogate of the overall mass-transfer coefficients. Gas-phase coefficients for H_2S are available in the literature;[4] the values for H_2Se are expected to be higher, but in our design, equivalent values are used as a conservative estimate.

Addition of sodium hypochlorite (NaOCl) in the caustic solution increases the rate of reaction and under certain conditions the overall scrubbing efficiency, but it forms highly soluble Na_2SO_4 which needs to be condensed before disposal.

Solutions of metal cations (e.g., $CuSO_4$) give another scrubbing liquid alternative which have the advantage of a very rapid chemical reaction with H_2Se and H_2S, to form easily precipitated salts. This option, however, was not investigated in our study because there is no industrial experience in using it.

Carbon Adsorption

H_2Se and H_2S are adsorbed and retained on the surface of activated carbon. Especially impregnated (IVP) carbon also reacts chemically with H_2S. The chemical reactions on carbon are not known, but a possible reaction of H_2S with O_2 forming S and H_2O has been reported.[5] H_2Se is expected to be chemically adsorbed by IVP carbon, but there are no data to substantiate this assumption. High concentrations of gases cannot be adsorbed on IVP carbon because the heat of adsorption is high[6] and the great amount of heat generated can reduce the bed capacity of adsorption. According to a manufacturer of IVP carbon, the material can handle gas steams with H_2S concentration up to 1000 ppm and can reduce it to below 4 ppb. IVP carbon can adsorb up to 20% of its weight in H_2S.

PROPOSED CONTROL SYSTEMS

A very important measure to reduce the impact of an accidental gas release is to keep only limited quantities of the gas in storage. We assumed a one-week inventory for each gas, stored in two cylinders, and as a worst reasonable scenario the release of the total contents of one H_2Se cylinder, hence 16 kg H_2Se.

The severity of a release can be reduced by flow-restricting valves within the gas cylinders to limit the leakage flow and therefore, allow time for engineering and administrative controls. A release can also be ameliorated by providing high purge flow rates for quick transport of the released gas to the treatment equipment. A gas line leak can be controlled by automatic shut-off valves activated by gas detectors and purging of released gas to the control equipment through race ways enclosing the pipes.

The proposed toxic gas control system for H_2Se and H_2S consists of three discrete components: (i) ventilation system, (ii) accidental release treatment facility and (iii) routine emissions treatment facility which may provide fine control to the gases leaving the accidental release systems.

We considered two treatment alternatives for reducing an accidental toxic gas release to an allowable emission level. Primary considerations were simplicity of design, control and operation that guarantees more effectiveness for hard to accurately describe accident conditions. Low maintenance requirements and low cost were also taken into account. One alternative is to treat all purged air continuously in a large venturi/packed-bed scrubber at a rate of 425 SCFM, and, in the event of an accidental release, to feed additional chemical to the system to neutralize the toxic gas.

The other alternative is containment of released gas and purged air in a custom-made closed-system scrubber, using a gas compressor and partial treatment of the gas in a closed system. The scrubber vessel should withhold modest level pressure (up to 10 atm) in order to contain the worst reasonable release and still be of practical size. The second stage of control is bleeding of the gas to the routine emissions treatment equipment to lower the concentrations to an allowable level.

Ventilation System

The first component of the toxic gas control system design is the ventilation system. It serves two major purposes: (i) to contain and divert accidental releases at the storage area to the accidental release treatment facility and (ii) to purge work areas where toxic gases might leak.

The ventilation system comprises air-purged ducts, which enclose all toxic gas lines, and air-purged cabinets, where the toxic gas cylinders are stored. Separate ducts are used for the gas lines entering and leaving the photovoltaic cell sputtering and metal chambers. The purged air for the unreacted gas lines (i.e., those leaving the photovoltaic manufacturing facility) is combined with the feed gas purge air after the latter exits the gas cabinets. The combined purged air is then sent to the accidental release treatment facility. Under normal conditions the combined air flow to this system is 200 SCFM which increases to 425 SCFM during an accidental release to provide for quick transport into the control system.

The photovoltaic plant work area is also continuously purged. A blower just upstream of the stack is used to pull air through the clean room at a flow rate of 2,000 SCFM. During normal operation clean room purged air discharged through the stack dilutes the gas stream from the routine emissions treatment facility.

Venturi/Packed-Bed Scrubber Alternative

Purged air flows to the accidental release treatment scrubber system and then through a carbon adsorption bed (Figure 1). The scrubber system comprises a multi baffle venturi which is designed to give a 60% efficiency and a packed tower designed for 99.5% efficiency. For the efficiencies to be as high as specified under accidental conditions the reactions have to be very fast so that treatment is only mass-transfer limited. This may be accomplished by adding more NaOCl in the scrubbing liquid which reacts quickly with both H_2Se and H_2S. The low precipitation rates of the products are a remedy rather than a problem in the accidental release scenario. It reduces the possibility of plugging of the packed tower and the increased volume of residual does not pose a problem since use of the system is not expected to be frequent. The gas out of the venturi is diluted by purged gas before it is fed into the packed tower. This provides better contact area and mixing times and will increase the practical efficiency of the packed tower.

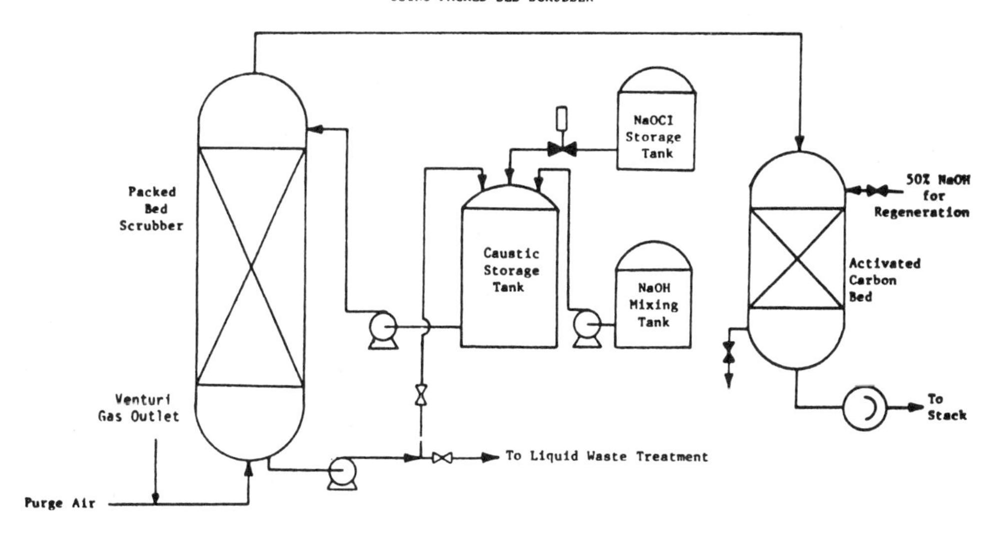

Figure 1. Accidental toxic gas release control using packed-bed scrubber

Under normal operation, no toxic gas will flow into the scrubber. The caustic recirculation rate (136 GPM) was set by the amount of NaOH needed to treat a worst case accidental release and can easily handle the worst possible line leak. A caustic make-up stream of 5.3 GPH is needed to neutralize the carbonic acid formed from continuously dissolving the CO_2 in the purged air into the scrubbing liquid. The total amount of 1 M NaOH in the recirculation tank (capacity 1500 gal) is about 500 gal.

If one of the gas sensors detects an accidental release, 500 gal of NaOCl are released into the recirculation tank. The amount of liquid available is enough to treat and keep in solution the toxic gas released. NaOCl is used to increase the reaction rate to the point where scrubbing is mass transfer limited. This will increase scrubbing efficiency in case of an accidental release. Since NaOCl has a short shelf life, it must be replaced monthly.

The scrubber itself is 1.5 ft in diameter and 12 ft high and should be constructed of material that is able to resist corrosion from NaOH under normal operation, and from NaOCl for at least the short time of operation during an accidental release. The packing is 25-mm polypropylene intalox saddles, which are resistant to corrosion by NaOH or NaOCl and have very good liquid redistribution properties. The concentration entering the scrubber during a worst case release would be 30 vol% initially and would be reduced to 12.8 vol% after 1 min (by the dilution effect of 500 ft^3 bulge in duct). This would result in 300 ppm leaving the packed-bed scrubber initially and would be less than 100 ppm after 1.5 min. Since this is still above the TLV, further treatment of the toxic gas is necessary.

An activated carbon adsorption bed 3 ft in diameter is used to reduce the concentration of toxic gas below the TLV. It is designed to treat 1075 SCFM of gas with a maximum toxic gas concentration of 300 ppm for short periods. Under normal operations, only purged air with no toxic gas will be flowing through the carbon bed. In case of an accidental release, the 300 ppm from the packed-bed can be treated by the carbon adsorption system to below 30 ppb initially (below TLV) and to below 4 ppb after 1.5 min. After 21 min, the concentration of H_2Se everywhere in the system will be below 4 ppb.

Containment and Scrubbing Vessel Alternative

Another alternative for the treatment of accidental releases of H_2Se and H_2S requires the use of a 1500 ft^3 containment vessel for temporary storage and partial treatment of the toxic gases. The vessel, as well as a 425 SCFM, 106 HP compressor and a 20 GPM, 3.3 HP pump, have a maximum operating pressure of 10 atm (see Figure 2).

During normal operation, purged air bypasses the containment vessel and flows at 200 SCFM directly to a 3-ft diameter activated carbon adsorption bed. If an accidental release is detected by a gas monitor, purge-air flow increases to 425 SCFM and is diverted to the containment vessel, which has venturi nozzles at its top through which 1 M NaOH is pumped from a 500 gal storage tank and scrubs the toxic gases. A level detector opens a drain when the caustic solution has reached a preset level. The caustic can be recirculated or treated as liquid waste. The efficiency of this control alternative is not known, but since it will operate in a closed loop for long times it can be assumed that it will treat the gases to the 14,500 ppm level (at 30% efficiency it will take about 4.5 hrs). When the toxic gas concentration inside the vessel falls below 14,500 ppm, gas is fed to the routine emissions treatment facility at a moderate flow rate of 5 SCFM. When the inlet toxic gas concentration drops below 100 ppm, the gas from the containment vessel is directed to the adsorption bed. This latter step accelerates the treatment time by making use of a larger carbon adsorption bed.

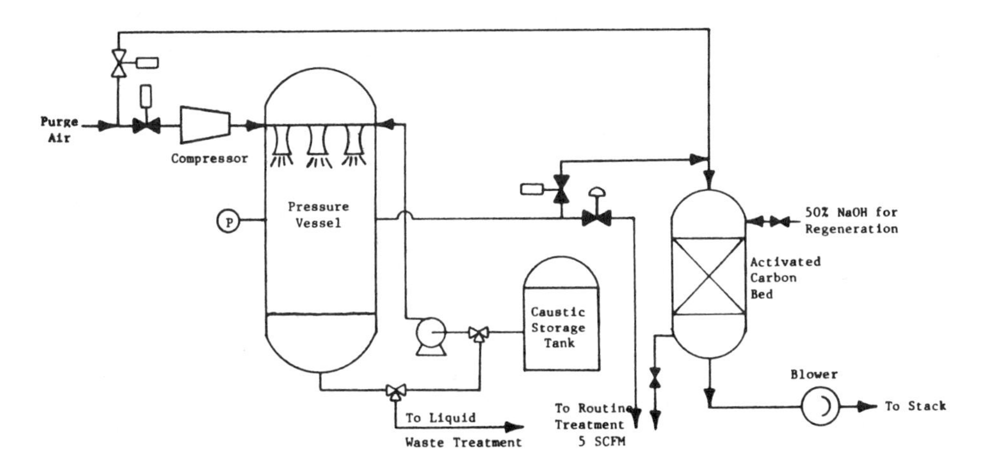

Figure 2. Accidental toxic gas release control using containment vessel

The routine emissions facility comprises a venturi scrubber, a packed-bed scrubber, and an activated carbon adsorption bed in series and can reduce the total toxic gas concentration from 14,500 ppm to 0.02 ppm. The off-gas is first contacted with 1 M NaOH in a venturi and spray scrubber of 60% efficiency where the effluent concentration reduces to 5800 ppm. The gas leaving the venturi/spray scrubber is fed into the bottom of a packed-bed scrubber designed for 99% scrubbing efficiency.[7] The gas leaving the packed-bed scrubber, which contains approximately 60 ppm toxic gas, is diluted with air to 10 SCFM before entering a carbon adsorption bed, which reduces the toxic gas concentration down to 4 ppb. This gas stream can be further diluted by a factor of 200 in the stack with purged air from the clean room to give a final discharge toxic gas concentration of only 0.02 ppb, which is three orders of magnitude below the allowable in-stack concentration for H_2Se.

Economic Analysis

The first component of the proposed design, the ventilation system, is standard equipment similar to that existing in any facility that uses toxic gases. Costs of this system and the ductwork and manufacturing enclosures were not calculated because they are considered process and not control equipment costs. Costs of additional ductwork and valving used in the routine and accidental treatment facilities are included in this economic analysis.

The results of economic analysis for the accidental release control alternatives are presented in Table II. Detailed control costs can be found elsewhere.[7] The estimated total annual cost of a toxic gas treatment system for a photovoltaic cell manufacturing plant is about 0.92¢/W_p for the scrubber accidental release alternative (excluding the cost of venturi), and 1.26¢/Wp for the containment alternative, assuming a plant life of 10 years and an interest rate of 15%.

A sensitivity study indicates that annual costs for both accidental release scenarios increase linearly with purge air flow rate, or the amount of H_2Se released. In both alternatives, the annual cost of accidental treatment only triples for an eight-fold increase in the amount of H_2Se released. Although it appears that the cost of the containment alternative is slightly higher than the scrubber alternative and increases more rapidly with purged air flow rate, a more detailed study of both alternatives should be completed before system selection.

The incremental costs per watt peak of electricity generating capacity is only about 0.1% of the $6/Wp production cost and 2% of the projected $1/Wp production cost. While the control system costs are only a small fraction of the total cost of producing $CuInSe_2$ photovoltaic cells, ways of reducing these costs even further need to be investigated in an effort to make photovoltaic cell production more competitive.

The Issue of Reliability

Are systems designed to control accidental gas releases going to function the way they are expected to do during a real emergency? The two suggested systems are based essentially on the same chemical system (e.g. NaOH and H_2Se) but on different operation modes (stand-by versus on-line system). This difference is especially important in comparatively examining the expected reliability of the two systems. The stand-by system needs to be triggered and successfully started upon an accident whereas the on-line system will be continuously operating. On the other hand the stand-by system will require less maintenance and repairs and therefore the corresponding times that the system will be unavailable will be shorter. Hazard analyses can aid in understanding the failure modes of these systems and evaluating their expected reliability. From the course of such analyses specific measures which can improve reliability (e.g. redundancy of critical components, optimization of testing and maintenance frequencies) can be determined.

CONCLUSIONS

This study shows that systems which could contain and control accidental releases of highly toxic H_2Se and H_2S in $CuInSe_2$ photovoltaic cell manufacturing facilities can be installed and operate at reasonable cost.

Two systems, a venturi/packed-bed and a containment/scrubbing system appear to be suitable for the control of these accidental releases. The total annualized cost of each of these systems is about $0.012/W_p$. The incremental cost of treatment increases with the amount of toxic gas accidentally released and is tripled for an eight-fold increase in the amount of gas released. These two proposed accidental release treatment alternatives should be examined in greater detail to determine which is expected to be more reliable. Also, the assumptions on chemical interactions of H_2Se with 1 M NaOH and H_2Se adsorption on activated carbon should be confirmed experimentally before implementing the proposed treatment systems.

REFERENCES

1. P.D. Moskowitz, V.M. Fthenakis and J. Lee, Potential Health and Safety Hazards Associated with the Production of Cadmium Telluride, Copper Indium Diselenide, and Zinc Phosphide Photovoltaic Cells, BNL 51832, Brookhaven National Laboratory, Upton NY (1985).

2. "New York State Air Guide-1, Guidelines for the Control of Toxic Ambient Air Contaminants," New York State Department of Environmental Conservation, 1985-1986 ed.

3. R.C, Reid, J.M. Prausnitz, and T.K. Sherwood, The Properties of Gases and Liquids, 3rd ed., McGraw-Hill, New York (1977).

4. R.H. Perry and D. Green (eds.), Chemical Engineering Handbook, McGraw-Hill, New York (1984).

5. W.D. Lovett, and F.T. Cunniff, "Air pollution control by activated carbon," Chemical Engineering Progress, 70 (5), 43 (1974).

6. J. Andrieu, and J.M. Smith, "Adsorption rates for sulfur dioxide and hydrogen sulfide in beds of activated carbon," AICHE, 27 (5), 840 (1981).

7. P.K. Fowler, D.G. Dobryn, and C.M. Lee, Control of Toxic Gas Release During the Production of Copper-Indium-Diselenide Photovoltaic Cells, BNL 38777, Brookhaven National Laboratory, Upton, NY (1986).

Table I. Toxic levels for H_2Se and H_2S.

Toxic gas	IDLH (ppm)[a]	TLV (ppm)[b]	AAL (ppm)[c]	Allowable in-stack conc. (ppm)
H_2Se	2	0.05	0.0002[d]	0.02
H_2S	300	10	0.01	1

[a] The IDLH level is defined as the maximum concentration of a toxic gas that a person can be subjected to for an exposure time of less than 30 min.
[b] The TLV is the maximum concentration that an employee working an 8 hr/day, 5 day/wk can be exposed to during a 30-yr work life without producing an observable adverse effect.
[c] AAL is the "acceptable ambient level" determined by the New York State Department of Environmental Conservation.
[d] The AAL for H_2Se was determined from the product of the TLV and a gas safety factor 1/300.

Table II. Economic analysis of toxic gas treatment system.

	Routine emissions treatment facility	Accidental release treatment facilities alternative	
		Scrubber	Containment
Total capital costs (1986 dollars)	$50,000	$290,000	$500,000
Annual operating costs	28,000	35,000	26,000
	Incremental Costs per Wp		
Amortized capital costs	0.10¢	0.57¢	1.00¢
Annual operating costs	0.28	0.35	0.26
Total incremental costs	0.38	0.92	1.26

PRACTICAL PERFORMANCE OF ENVIRONMENTAL CONTROL EQUIPMENT FOR SILANE, PHOSPHINE, AND DIBORANE FEEDSTOCK PROCESSING

Ashok K. Khanna and Raj Gupta
ARCO Solar, Inc., Chatsworth, California 91313

ABSTRACT

Due to the hazardous nature of the feedstock gases often found in thin film silicon:hydrogen alloy ("amorphous silicon") deposition, it is important to control and render harmless those reactants that are not consumed as inert coating. Environmental control equipment used for such purposes must meet important safety, throughput, efficiency, cost, and reliability requirements. This review discusses the advantages and disadvantages of scrubbing and incineration methods and actual performance results obtained for such systems.

INTRODUCTION

Because of the quantities of silane, phosphine, and diborane used in photovoltaic (PV) manufacturing, the emission flows must be efficiently minimized to an acceptable level, and the reactants that are not converted into product must be transformed into a benign form.

An environmental control system can be designed that prevents contaminants from attaining concentrations that may be of concern to the health of people or protection of the environment. Control at the source provides an important opportunity to both minimize the complexity and reduce cost. Proper selection of environmental control equipment requires detailed information about the performance and applicability of the hardware and the process streams under a variety of conditions. Performance monitors are important to assure that the equipment is operating efficiently during all phases of processing by tracking variables such as composition, flow rate, impurities, and variations of flows.

Another major consideration is abatement at the source versus a remotely located system. The process designer must be alert to any complex by-products and second level issues such as particulate buildup in packing material that can affect efficiency over time. A final critical factor is equipment maintainability, which should include some evaluation of skill level required to support the equipment.

It is important to note that the procedures and results described here may not be appropriate in all processing operations using the same or similar materials.

WARNING: Information in this report is furnished without warranty expressed or implied. No use of this information should be made without first obtaining the assistance of scientists, engineers, and professionals who have specialized training and experience in the use and handling of hazardous toxic materials. The user of this information assumes any and all responsibility for its own use of this information and expressly relieves ARCO Solar, Inc., its parent, subsidiary and affiliated companies of any and all responsibility or liability for injury to property or persons including death and any loss, damage, or expense arising out of or in any way connected with such information.

EQUIPMENT STUDIED

In the basic arrangement for controlling silane and dopant by-products from a thin film silicon deposition system, gases leaving the process chamber/vacuum pumps are diluted with nitrogen to maintain the silane concentration at or below 1% by volume; this reduces fire and/or explosion hazards. Vacuum pumps are continuously purged with nitrogen to avoid any trapping of silane and to minimize buildup of silicon dioxide and accumulation of dopant gases dissolved in the pump oil. This type of basic system design has resulted in trouble-free operations.

This paper discusses the performance of scrubbers and incinerators, plus a bag house (Fig. 1). Results of phosphine chemisorption are detailed elsewhere.[1]

The composition of the effluent stream fed to each system under study (except as mentioned) is given in Table I. All systems were operated at 5-7 inches of negative water pressure.

RESULTS AND DISCUSSION

A gas sampling system designed primarily for impurity analysis of process gases was used to characterize performance of the environmental control units. This system is connected to three gas chromatographs and provided flexibility in performing a wide variety of gas analyses at relatively low concentration levels. Appropriate filters were used between the sampling line and a control unit to avoid introduction of any particulates in the gas stream for analysis. The sampling lines were purged with nitrogen to remove any buildup of moisture prior to gas sampling. The sampling line must be cleaned and made airtight to maintain the integrity of the gas stream for accurate evaluation of a control unit. Excessive moisture in the line would partly trap some hydrides, B_2H_6 in particular, by causing hydrolysis. Additionally, the sampling system must be checked carefully for air leaks to avoid any safety hazard.

Details of the gas sampling and analytical techniques are described elsewhere.[2] The results for the performance evaluation of the wet scrubbers and incinerators are summarized in Table II.

WET LIQUID SCRUBBERS

Scrubber design requires knowledge of the process, including information concerning:

1. Inlet gas volume, temperature, humidity, and pressure;
2. Peak gas and particulate loading;
3. Average gas and particulate loading;
4. Scrubbing liquid and its chemistry with gas as a function of solution life;
5. Particulate size distribution;
6. Contaminants and means of disposal.

Gas-liquid absorption (wet scrubbing) is a mass transfer process where the pollutant gas is transferred from the gas stream into the liquid stream. The rate of mass transfer is dependent on the gas-liquid interface area, the differences in concentration of the two phases, and the chemical species present.

Scrubbing that uses liquids for gas absorption relies on creating large

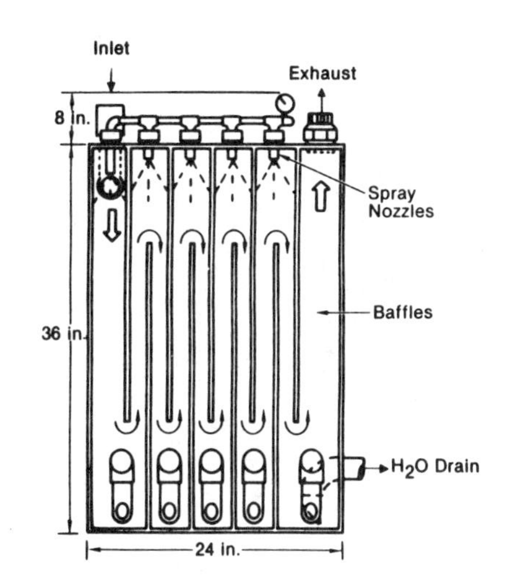

Wet liquid scrubber with spray/baffle chamber.

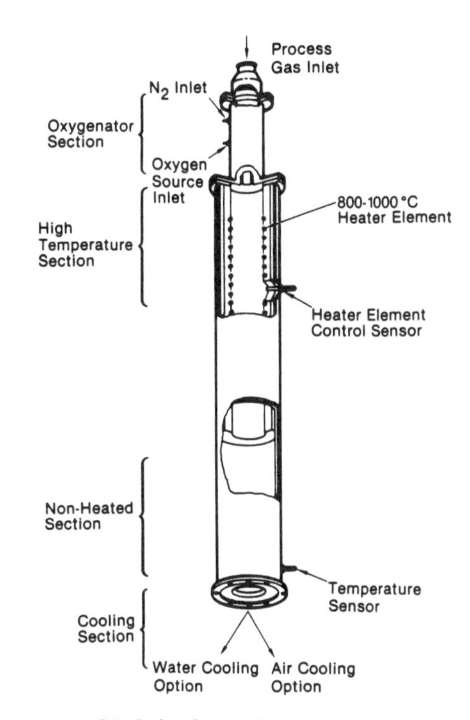

CDO incineration unit.

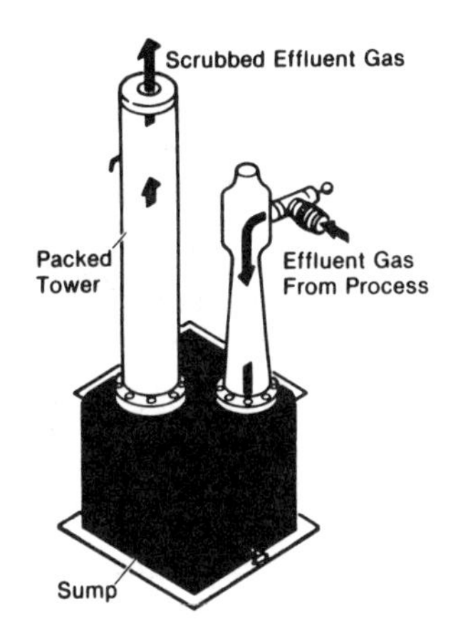

Wet liquid scrubber with venturi followed by packed tower.

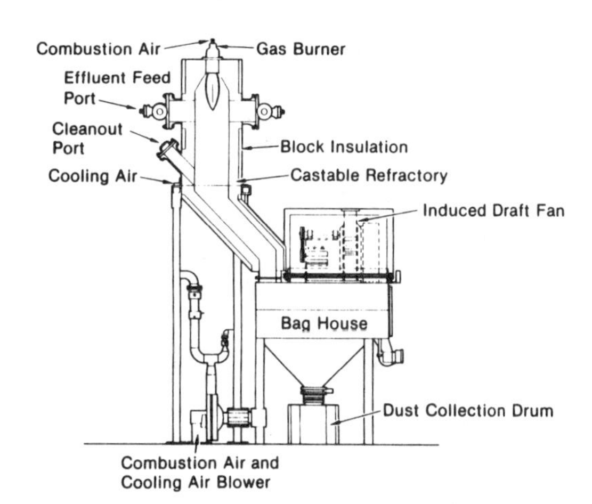

Gas fired incineration unit.

Fig. 1. Control systems evaluated for study.

Table I. Effluent stream fed to control systems.

Gases	Flow Rate (per minute)
Silane (100%)	2.0 liters
Nitrogen	100.0-200.0 liters
Phosphine (<2%)	~200 cm^3
Diborane (<2%)	~150 cm^3
Pump Oil Vapors	trace levels

Table II. Efficiency of wet scrubbers vs incineration in removal of effluent gases.

	Silane	Phosphine	Diborane
Wet Scrubber (NaOH) Spray Chamber			
Inlet (ppm)	170	9.9	hydrolyzed
Outlet (ppm)	148	9.5	~0
Eff. (%)	13	4	N/A
Wet Scrubber (NaOH) Venturi/Packed Column			
Inlet (ppm)	7180	not done	not done
Outlet (ppm)	5150	not done	not done
Eff. (%)	20	not done	not done
Thermal Incinerator CDO Unit - 1000°C			
Inlet (ppm)	2036	29	108
Outlet (ppm)	<1	29	<2
Eff. (%)	~100	~0	>98
Thermal Incinerator Gas Fired Unit - 815°C			
Inlet (ppm)	5000	35	7.5
Outlet (ppm)	<4.4*	<0.21*	<0.003*
Eff. (%)	>98	>92	>99

*Detection Limit

liquid surface areas. The techniques to create these large surface areas can be divided in two basic categories: those which flow the liquid over a media (packing, meshes, etc.) and those which generate a spray of droplets.

Spray Scrubber

The wet liquid scrubber with spray/baffle chambers provides a simple and basic form of spray scrubbing. The gas velocity is reduced in an expansion area where it comes in contact with liquid sprays that provide a reasonable gas-liquid interface for reaction. Low gas velocity helps large particulates fall to the bottom of the chamber.

The liquid used is 75 kg of approximately 10% NaOH in water. The solution was monitored weekly for pH, silicate concentration, and total dissolved solids. The system is a total recirculating type with intermittent addition of make-up water to maintain the desired liquid level.

These gases may be partially hydrolyzed in solution in the presence of hydroxyl ions. The expected chemical reactions are mentioned by Khanna.[1]

Known flows of silane, phosphine, and diborane were introduced into the scrubber through a processing system. The results indicate low scrubbing efficiency of the system in the removal of silane and phosphine under the particular operating conditions described because phosphine does not significantly react with caustic solutions. The scrubbing efficiency of diborane is expected to be high due to its rapid hydrolysis in the scrubber solution. Diborane was not detected at the scrubber inlet, probably due to the presence of moisture in the scrubber duct or absorption in pump oil. Since phosphine is sparingly soluble in the scrubber solution being used, it does not impart any acidity or alkalinity. Silane scrubbing efficiency depends on effluent stream composition, scrubber design, and equipment maintenance over time.

The slow rate of reaction responsible for transforming these gases into benign form is possibly one reason for poor scrubbing performance of the system. Secondly, the low solubility of the effluent gases in the scrubber solution (already saturated by high solids) decreases the contact time between the reactants necessary for complete reaction. Finally, poor mixing and relatively low wet surface area would result in poor mass transfer coefficients, hence inefficient scrubbing.

The levels of diborane and phosphine in the gas stream coming from the system (processing chamber) bypass were measured much lower than would be expected from the initial concentrations of gases originally fed into the stream. The gases through the bypass are run by the mechanical pump that is backed by a roots blower.

Pump oil can be considered a possible source for trapping these gases to some extent by absorption/chemisorption phenomena. A systematic study was performed to determine if pump oil removes any fraction of diborane (0.3%) or phosphine (1%) from the gas stream. No diborane was detected, suggesting that it possibly was hydrolyzed by the moisture in the pump oil. It was also established that 60-65% phosphine is absorbed in the pump oil. The chemistry of phosphine absorption is not well characterized. The results were fairly repeatable during the experiments conducted over a period of 1 hour in each case.

Samples of oil from the deposition process pump were drawn and analyzed for phosphorous and diborane. No such elements were observed in the samples, suggesting that they possibly were precipitated out from the oil. Further studies are required to confirm the potential absorption or conversion into other components of diborane and phosphine in pump oil.

Venturi Scrubber

The wet liquid scrubber with a venturi eductor followed by a packed column used a recirculated scrubbing solution consisting of 45 kg of 7% NaOH in water. A known flow of silane was introduced to the converging section of the venturi eductor then diverted to a packed column. The results indicate that the venturi eductor is 15% efficient in decomposing silane. A combination of venturi eductor and packed column is 20% efficient. No change in efficiency was noticed when the flow rate of scrubbing solution in the venturi eductor was reduced by two-thirds.

Our experience has shown that the wet liquid scrubber system is maintenance intensive because of the large amount of sodium silicate generated and the abrasive nature of silicates in solution. Silane will preferentially react with any oxygen in the system and form SiO_2. Based on material balance calculations and the above flow rates, a 1.5 kg quantity of silane reduces the NaOH concentration in the scrubber to 5% while the total solid concentration reaches about 13.8%. At this point, the scrubber solution would be drained and the scrubber cleaned and replenished with 75 kg of fresh 10% NaOH. In so doing, about 80 liters of liquid waste has been generated which now creates a secondary waste disposal problem. Potassium salts of silicates, phosphates, and borates are more soluble than their sodium salts. Thus, potassium hydroxide as a scrubbing solution has more capacity to hold solids and thus will function longer before it must be replenished.

An additional problem is that if the solution is not monitored weekly, SiO_2 will accumulate and potentially damage both the pump impellers and spray nozzles.

Since caustic solution systems are inefficient for phosphine scrubbing, a second stage wet scrubber is required. An oxidizing medium scrubber solution (e.g., potassium permanganate or sodium hypochlorite solution) can be used to remove phosphine. But the processor must be sure that most of the silane is eliminated before the effluent reaches second stage scrubbing. Silane can form an <u>explosive</u> mixture with oxidizing medium scrubber solutions. System safety can be achieved by monitoring silane between two stages to trigger a second stage bypass in case the first stage scrubber fails.

THERMAL INCINERATORS

The basic types of waste gas incineration systems are indirect heat, direct heat, and catalytic. Catalytic converters are not used since the self-sustaining reaction of SiH_4 does not require a catalyst, and no information was found in the literature regarding a catalytic system to improve the oxidation rate of phosphine.

CDO Unit

Controlled combustion, decomposition, and oxidation (CDO) is a gaseous effluent conditioning technique that uses indirect heat and reduces hazards associated with flammable and hazardous gases and vapors.[3] This technique may also prove to be relatively effective in case of an accidental large release.

Understanding the combustion kinetics of effluent gases and their dependence on process conditions is important. Process parameters such as temperature, concentration of reacting species, residence time, mixing rate of fuel/oxidant, and percent accessible oxygen dictate how complete the reaction will be.

Time, temperature, turbulence, and oxygen are required for combustion. Each of these variables must be optimized for good combustion. The temperature should be high enough to oxidize the contaminants, which is typically much higher than the autoignition temperature. The correct temperature is also essential for controlling the generation of nitrogen oxides. It may be noted that long residence times dictate the need for larger control units and the increased temperatures require more heat and place additional limitations on materials of construction. Since good turbulence is the easiest of the four factors to supply, it should always be maximized.

Thermal incinerators use gas or electricity to maintain the control unit temperature at ~1000°C in order to sustain combustion. Since electrically heated units do not require air for natural gas combustion, these units tend to be smaller in size and run cleaner. Contaminants such as sulfur in natural gas are not introduced. With gas combustion, attention must be paid to flame temperature to avoid formation of nitrogen oxides.

Process exhaust gases are mixed with controlled amounts of oxygen and air in the oxygenator section of the CDO. The inlet of the CDO unit is designed to allow for the free flow of the gases into the system with minimum back pressure. The oxygenator is specifically designed so that the mixing occurs downstream of the oxygenator nozzles and near the beginning of the high temperature section. This section is maintained at ~800-1000°C.

The gas temperature is reduced in the non-heated section by mixing with cooling air. The temperature in this section is constantly monitored. Silicon dioxide dust is entrained in air and is filtered in a bag house. An effective method for handling large volumes of dust at flow rates >2.0 SLM (standard liters per minute) of silane is required or the CDO unit may plug in less than 15 minutes of operation.

The CDO approach has been shown to be highly efficient in the controlled combustion of silane and diborane, the results suggesting performance efficiency approaching 100% for silane. However, the system under the defined operating conditions appears less efficient in the removal of phosphine. It has been shown that higher residence time of the effluent in the hot zone may help to enhance the performance of the system. Under present conditions, the residence time for gases is approximately 0.5 seconds, which is lower than the suggested value of 2-4 seconds.

Silane combustion studies (Fig. 2) indicate that silane conversion efficiencies close to 100% can be attained if the hot zone is maintained at a temperature higher than 850°C. Burning efficiency drops at lower temperatures.

The effect of residence time on combustion efficiency of phosphine was studied at 950°C. Residence time for feed gases was changed by changing phosphine and purge nitrogen flow rates. As indicated in Fig. 3, a minimum of 3 seconds is needed to oxidize phosphine effectively. Although higher temperatures may improve system performance, equipment limitations did not allow operation of the unit above 1000°C.

The CDO system has an inherent advantage over wet scrubbing. It generates solid waste that is smaller in volume and easier to handle and may generally be disposed of as non-hazardous waste. Also, there is no commitment to long-term containment such as for liquid waste.

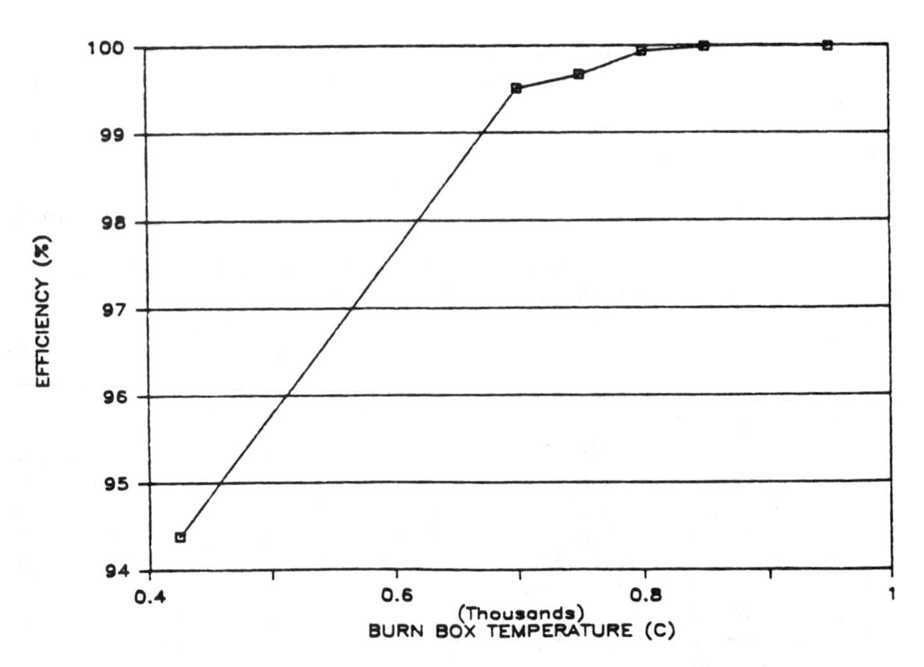

Fig. 2. Silane combustion in CDO unit.

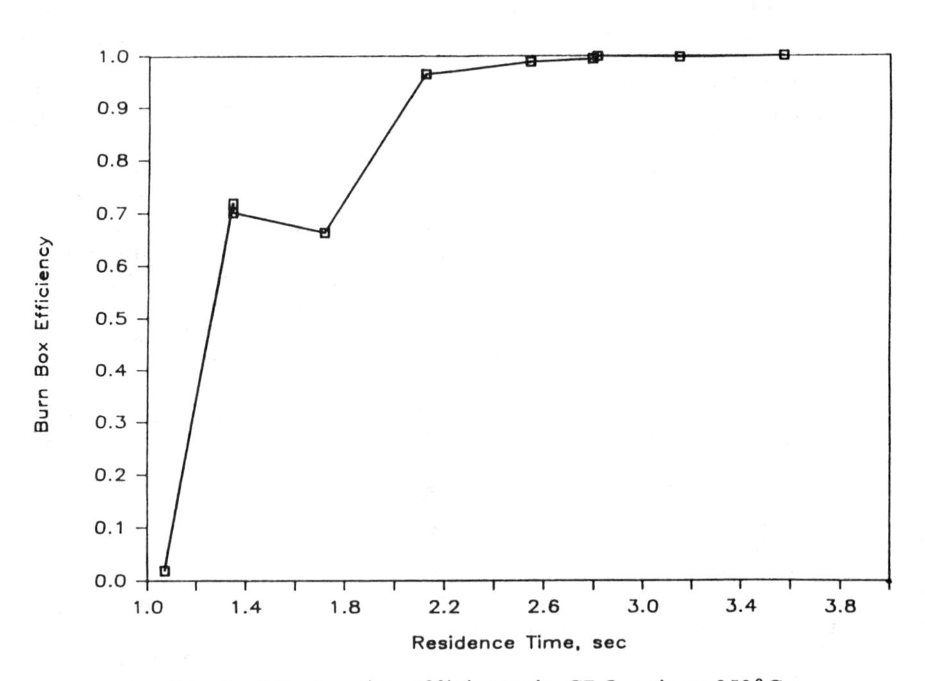

Fig. 3. PH_3 burning efficiency in CDO unit at 950°C.

Gas Fired Unit

The gas fired direct flame incineration method is normally used with materials that are at or near their lower combustible limit when mixed with air. Natural gas or auxiliary fuel is added to sustain combustion in the burner.

The unit includes incinerator chamber, gas fired combustion system, air cooling system, bag house, induced draft fan, and controls. The incinerator is gas fired, and the process effluent stream is fed radially and is directly injected and drawn into the flame. Flame and high air flow provide tremendous turbulence and mixing of the effluents thus promoting complete combustion.

The incinerator shell is made of 3/16 in. carbon steel plate with castable refractory insulation which reduces thermal losses. The unit is maintained at 815°C, and gases have a residence time of about 10 seconds.

The gases are cooled by air to below 150°C and fed into a bag house to filter out SiO_2 dust. The bag house must be sized for about 1:1 air to surface area ratio to ensure trouble-free operation. Bags are made of fire retardant material with a slick finish. The dust is collected in a drum under the bag house. The results in Table II show no detectable silane, phosphine, or diborane.

BAG HOUSE

Following incineration, a bag house removes SiO_2 dust particles from the gas stream. The major collection occurs when particles build up on the filter surfaces and bridge the pore openings, which are often larger than the particles themselves. The accumulation of dust becomes a "filter cake" which now acts as an efficient filter medium.

The pressure drop across a filter depends upon the flow rate, the porosity of the filter medium, and the cake. The pressure drop across a clean cloth increases for the first few cycles then stabilizes. Collection efficiencies of filters can be high, but particles 0.1-1.0 microns in size are difficult to remove. Gas velocities through the filter medium depend on filter fabrics and typically are rated for 10-30 cfm per ft^2 filter area at negative pressure of 0.5 inches of water.[4]

Equipment reliability is a significant issue due to the high volume of fine SiO_2 particulates generated. The density of SiO_2 dust is low (0.0275 gm/cm^3). The combustion and decomposition of 1.0 kg of silane will produce 1.875 kg of SiO_2 which will occupy approximately 70 liters of space.

The equipment necessary to handle dust effectively depends mostly on the particulate size. The SiO_2 particle size will be affected by the amount of nitrogen dilution.

Special filter bags must be used to operate the systems continuously. Safety may dictate the need for filter bags made of "Nomex" polycarbonate fibers due to the potential rise in temperature of the gas flow as it enters the bag house. These bags are rated for higher temperature but require lower gas flow rates. Due to the weave and texture of the material, these bags do not shed SiO_2 well and plug quickly. Our experience shows that the bag house systems work more efficiently with bags made of "Gore-Tex" filter fibers. These are Teflon-coated fibers rated for high temperature service. Bags made from this material filter the dust efficiently without any bleed through and also shed the particulates more readily due to the smooth, slick surface of the fibers.

The fine particle size dictates that protective equipment (e.g., dust masks) should be used by workers while handling SiO_2 dust.[5]

CONCLUSIONS

From the above discussions, it can be concluded that hazardous materials involved in the manufacturing of certain photovoltaic products can be controlled in an environmentally safe manner. Wet liquid scrubbers were found to be relatively inefficient in decomposing those reactants that are not consumed in the silicon coating process. Additionally, scrubbers pose a problem of secondary disposal of liquid waste. In contrast, flame incinerators and CDO units are 99.9+% efficient in removing silane from the effluent stream if the temperature is maintained above 850°C. Our experience indicates that phosphine can be destroyed in flame incinerators and CDO units when the residence time is >3 seconds.

For pilot research scale operations, properly designed and sized CDO systems can be effective. For large scale operations, natural gas fired incinerators appear to be the most efficient for effluent control for thin film silicon photovoltaic manufacturing operations since this type of equipment is commercially available and can be easily scaled. Proper sizing of the bag house or other dust collection tool and selection of filter material are critical to ensure continuous operation.

Concern for community and employee health and well being and control of hazardous agents in any manufacturing operation are important. Environmental control equipment must perform reliably and effectively. It is important to demonstrate and routinely monitor the control equipment efficiency using analytical techniques appropriate to individual circumstances.

ACKNOWLEDGMENTS

This work was funded by the Biomedical and Environmental Assessment Division, Department of Applied Science, Brookhaven National Laboratory, under contract to the Photovoltaic Energy Technology Division, Conservation and Renewable Energy, U.S. Department of Energy. Grateful acknowledgment is extended to P.D. Moskowitz, who served as Project Officer, and to V.M. Fthenakis for their guidance and comments. Special thanks are due to I. Littorin of ARCO Solar for her encouragement and leadership and to M.E. Vojtek, A. Joseph, N. Singh, and D. Houk for their contributions.

REFERENCES

1. Ashok K. Khanna et al., *Evaluation of Hazards Associated with the Manufacture of Thin Film Solar Cells,* technical report prepared for Brookhaven National Laboratory, Upton, N.Y., under contract 303575-S (Jan. 1987).
2. Ashok K. Khanna and Raj Gupta, *Solar Cells* 19, 301 (1986-1987).
3. "Exhaust Gas Conditioning Systems," Technical Release, Innovative Engineering (San Jose, Calif., 1983).
4. H. Hesketh, *Fine Particles in Gaseous Media,* 2nd ed. (Lewis Publishers, Chelsea, Mich.).
5. P.D. Moskowitz et al., *Health and Environmental Effects Document for Photovoltaic Energy Systems: 1983,* BNL 51676, Brookhaven National

EFFICIENTLY HANDLING EFFLUENT GASES THROUGH CHEMICAL SCRUBBING

Tim Herman Scott Soden
Airprotek Inc., Napa, CA 94558

ABSTRACT

This paper is presented as an information source for efficiencies of chemical scrubbing. In it, we will discuss the specific problems of scrubbing silane, disilane, diborane, phosphine, hydrogen selenide and arsine. We will explain the scrubber dynamics, gases and flow rates used along with liquid mediums. The equipment and procedures used for testing, as well as the determination of the results, will be discussed. We intend to give examples of possible reactions and documentation of our efficiencies. Installation and maintenance will be touched, as well as our experiments into accidental catastrophic releases. From all of this we will derive conclusions as to the best possible means of wet chemical scrubbing.

INTRODUCTION

The photovoltaic and semiconductor industries require the use of corrosive, toxic and/or pyrophoric gases in their manufacturing processes. These gases, if left untreated, can create severe hazardous conditions which can affect both personnel and equipment, as well as environmental concerns.

The proper treatment of these gases is required in order to assure the safety and well being of personnel, equipment and the surrounding environment.

PROCESS GASES

There are many gases used in the various steps of manufacturing the final product. Due to the variations in manufacturing techniques, the process parameters are different for most companies. Concentration and flow rates for the various gases as well as process time will differ. These parameters, however, play a major part in chemical scrubbing.

Various processes will use different gases in the same flow stream making it more difficult to react those gases with a common medium. It is imperative that the gas treatment equipment supplier know all of the parameters of the process in order to properly suggest a treatment method.

In some cases the gases used in the process will react with water which will eliminate its potential hazard. This is the case with diborane (B_2H_6) [1,2]. which will form boric acid:

$$B_2H_6 + 6H_2O = 2H_3BO_3 + 6H_2 \quad (1)$$

Disilane will react with water as well to yeild silicon dioxide and hydrogen:

$$Si_2H_6 + 4H_2O = 2SiO_2 + 7H_2 \quad (2)$$

Hydrogen selenide also reacts with water producing a selenious acid;

$$H_2Se + 3H_2O = H_2SeO_3 + 3H_2 \quad (3)$$

These reactions have been proven many times in the field. However, we are not as fortunate with the other gases since they require more complex solutions for proper treatment.

TABLE I PROCESS GAS DATA [1,2,3]

GAS	FORMULA	TLV	IDHL	DOT HAZARDOUS CLASS
Arsine	AsH_3	0.005ppm	6ppm	Poison-Flammable Gas
Diborane	B_2H_6	0.1 ppm	10ppm	Poison-Flammable Gas
Disilane	Si_2H_6	-	-	Poison-Flammable Gas
Hydrogen-Selenide	H_2Se	0.2ppm	2ppm	Poison-Flammable Gas
Phosphine	PH_3	0.3ppm	200ppm	Poison-Flammable Gas
Silane	SiH_4	5.0ppm	-	Poison-Flammable Gas

In looking at the hydrides, AsH_3, PH_3, and SiH_4 we find that these are not reactive with water but will react with other media. First, we will look at AsH_3. AsH_3 is spontaneously inflammable and in contact with air burns to form white fumes of the trioxide:

$$2AsH_3 + 3O_2 = 3H_2O + As_2O_3 \quad [5] \quad (4)$$

This oxide in itself is still hazardous in a fine white powder and must then be scrubbed.

Various other data indicates that AsH_3 reacts with several different chemicals. An experiment performed and published in 1908 [6] indicates that AsH_3 will react with iodic acid (HIO_3):

$$6HIO_3 + 5AsH_3 = 5H_3AsO_3 + 3I_2 + 3H_2O \quad [6] \quad (5)$$

A reaction with potassium permanganate gives dipotassium hydrogen arsenate, manganic oxide and water which stay in solution;

$$AsH_3 + 2KMnO_4 = K_2HAsO_4 + Mn_2O_3 + H_2 \quad [7] \quad (6)$$

we can also react it with potassium hydroxide to form a solid potassium arsenide:

$$AsH_3 + 3KOH = K_3As + 3H_2O \quad [8] \quad (7)$$

TABLE VIII: SILANE EFFICIENCY TEST

SCRUBBER EFFICIENCY TEST

TEST GAS: SiH4 TEST PROCEDURE: GC HWD DATE: 10/28/86

TEST MODEL: SD101B SCRUBBING MEDIUM: 2% NAOH IN H2O

TEST #	GAS	FLOW sccm	CARRIER sccm	FLOW sccm	INLET CONC %	OUTLET CONC %	OUTLET PPM	EFFICIENCY %
1	SiH4	100	N2	1000	14.65	NR *	< 0.5	99.99+
2	SiH4	100	N2	1000	13.82	NR *	< 0.5	99.99+
3	SiH4	100	N2	1000	13.71	NR *	< 0.5	99.99+
4								
5	SiH4	200	N2	1000	24.67	NR *	< 0.5	99.99+
6	SiH4	200	N2	1000	24.59	NR *	< 0.5	99.99+
7								
8	SiH4	1000	N2	30000	1.13	.015	152.66	98.05
9	SiH4	1000	N2	30000	1.09	.047	471.69	95.91
10	SiH4	1000	N2	30000	1.14	.0039	39.64	99.65
11								
12								

* NR - No Reading
Sensitivity of HWD = 0.5 parts per million
Sensitivity of PID = 1.0 part per billion

TABLE IX: SILANE EFFICIENCY TEST

SCRUBBER EFFICIENCY TEST

TEST GAS: SiH4 TEST PROCEDURE: GC HWD DATE: 10/28/86

TEST MODEL: SD101B SCRUBBING MEDIUM: 2% NAOH IN H2O

TEST #	GAS	FLOW sccm	CARRIER sccm	FLOW sccm	INLET CONC %	OUTLET CONC %	OUTLET PPM	EFFICIENCY %
1	SiH4	1000	N2	10000	11.01	.824	8243	92.51
2	SiH4	1000	N2	10000	12.21	.858	8581	92.97
3								
4	SiH4	2000	N2	10000	19.96	1.329	13297	93.34
5	SiH4	2000	N2	10000	21.97	1.102	11201	94.98
6	SiH4	2000	N2	10000	19.74	1.059	10591	94.63
7								
8	SiH4	2000	N2	15000	13.52	.918	9188	93.21
9	SiH4	2000	N2	15000	16.96	1.318	13180	92.22
10	SiH4	2000	N2	15000	15.31	.991	9913	93.48
11	SiH4	3000	N2	10000	23.93	4.939	49390	79.36
12	SIH4	170	N2	60000	.2509	.145	1453	42.08

$$SiH_4 + 2H_2O = SiO_2 + 4H_2 \tag{11}$$

Testing shows that this is not at all true. SiH_4 is shown to dissolve in a solution of potassium hydroxide [11] and rapidly hydrolizes to hydrated silica [12] in the presence of a base. Formulating ideas from this data, additional tests have been done running SiH_4 and using aqueous NaOH as a reactive medium. The chemical reaction is believed to be:

$$SiH_4 + 4NaOH = NaSiO_4 + 4H_2 \tag{12}$$

The testing that we performed showed a clear view of what is necessary to scrub SiH_4. The tests were run using a model SD 101 R gas effluent treatment system which utilizes a flow/counter flow technique. Concentrations and flow rates are shown on Tables VIII and IX. The scrubbing system is similar in design to that in Figure 1.

FIGURE 1: AIRPROTEK INC. MODEL SD I R
GAS EFFLUENT TREATMENT SYSTEM

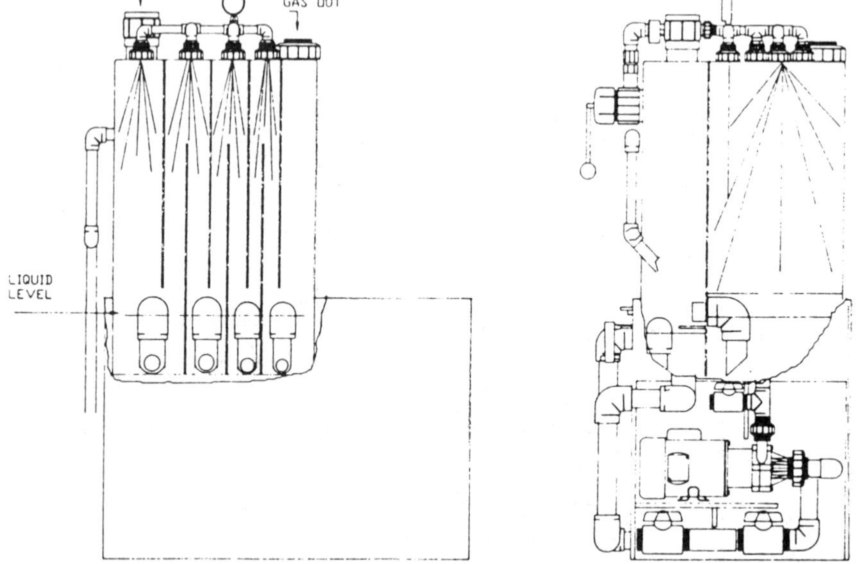

The inlet and outlet concentrations were measured utilizing a Perkin-Elmer Sigma 2000 gas chromatograph using a thermal conductivity detector. Results were printed out by a Perkin-Elmer LCI 100 integration system.

TABLE VI: INLET CONCENTRATION OF EFFICIENCY TEST FOR PH_3 USING NaOH

FILE 30 RUN 1 STARTED 15:30.5 86/01/30 PH3/NAOH TEST XS METHOD 1
PH3/KMNO4 LAST EDITED 15:30.5 86.01/30 SMP AMT 1.0

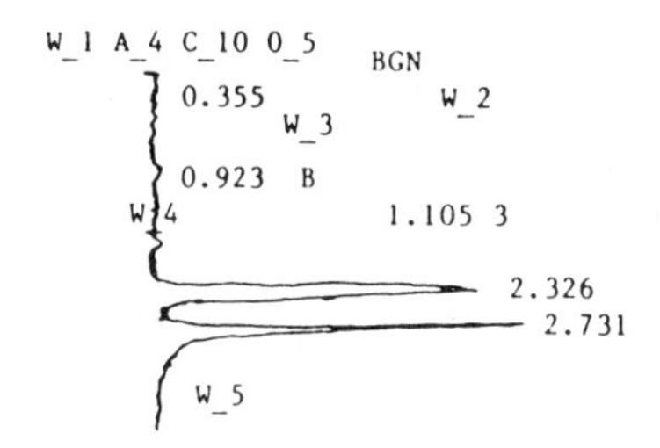

FILE 30 RUN 1 STARTED 15:30.6 86/01/30 PH3/NAOH TEST XS METHOD 1
PH3/KMNO LAST EDITED 15:30.5 86.01/30 SMP 1.0

RT	AREA	BC	RT/10	RF	PPM	NAME
2.731	54139		0.273	3.151926E-03	170.6421	PH3

2 MATCHED COMPONENTS 99.89% OF TOTAL AREA
3 UNKNOWN PEAKS UNRET PK 0.11% OF TOTAL AREA
5 PEAKS AREA REJECT 126535 TOTAL AREA

TABLE VII: OUTLET CONCENTRATION OF EFFICIENCY TEST FOR PH_3 USING NaOH

FILE 33 RUN 4 STARTED 15:40.2 86/01/30 PH3/NAOH TEST XS METHOD 1
PH3/KNMO4 LAST EDITED 15:30.5 86/01/30 SMP AMT 1.0

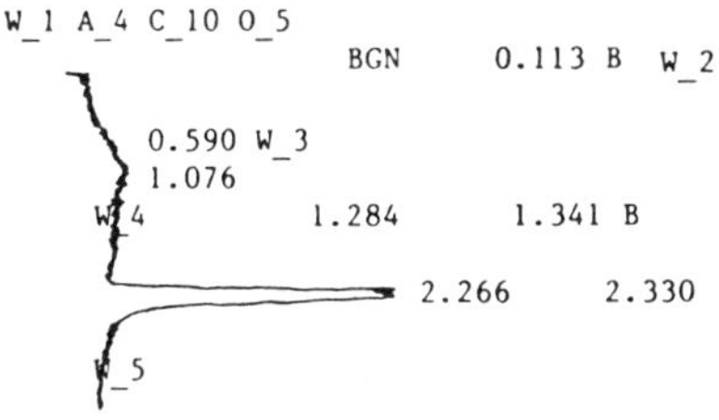

FILE 33 RUN 4 STARTED 15:42.2 86/01/30 PHS/NAOH TEST XS METHOD 1
PH3/KMNO4 LAST EDITED 15:30.5 86/01/30 SMP AMT 1.0

RT	AREA	BC	RT/10	RF	PPM	NAME
			0.274	3.151926E-03		PH3

2 MATCHED COMPONENTS 99.86% OF TOTAL AREA
5 UNKNOWN PEAKS UNRET PK 0.64% OF TOTAL AREA
7 PEAKS AREA REJECT 61405 TOTAL AREA

Silane is a gas that is commonly used in the industry, although very little is known about it. Recent testing has shown that it can be spontaneously flammable or collect and reach explosive conditions resulting in severe damage to life and property. It is widely believed that SiH_4 is soluble in water producing silicon dioxide:

TABLE III: LIQUID SCRUBBER EFFICIENCIES OF AsH_3 USING A NaOCL/KOH SOLUTION

TEST # 2

Test Gas: AsH_3 Test Liquid: 1%NaOCl/1%KOH in H_2O by weight
Concentration: 1.5%inN_2 (15000ppm)Liquid Volume: 25 gal.
Total Flow: 10 slm Liquid Flow: 2.25 gpm
Test Equipment: Perkin-Elmer Sigma 2000 Gas Chromatagraph, LCI 100 Integration System
Calibration: H.W.D.-150000 ppm in N_2

Test System: After scrubbing, the outlet of the scrubber was tested with a Perkin-Elmer Sigma 2000 Gas Chromatagraph first utilizing a hot wire detection system.

GRAPH FROM H.W.D. SYSTEM

File 13 Run 4 Stared 05:30.2 86/01/04 AsH_3 NaOCl/KOH XS Method 2
AsH_3 Conc. Out. Last Edited 02:05.5 86/01/02 SMP AMT 1.0

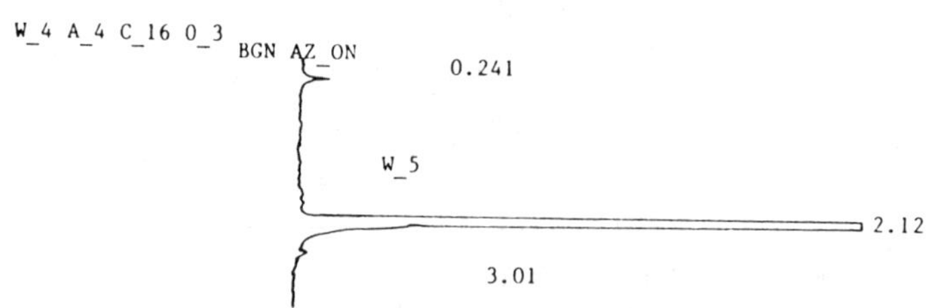

File 13 Run 4 Started 05:30.2 86/01/04 AsH_3 NaOCl/KOH XS Method 2
AsH_3 Conc. Out. Last Edited 02:05.5 86/01/02

RT	AREA	BC	RRT	RF	PPM	NAME
2.12	2029217E		0.702	1.808008E+08	2029217E.0880	Nitrogen
3.01	2793		1.000	4.124420E-02	113.1744	Arsine

2 Matched Components 99.97% of total Area
1 Unknown Peak Unret PK 0.03% of total Area
3 Peaks Area Reject 20388768 Total Area

115.1744 ppm equates to 99.3% efficient

Although NaOCL/KOH is an adequate scrubbing medium, it is apparent that $KMnO_4$ is much more effective at scrubbing. One observation made through the various tests was that the longer the gas was in contact with the liquid the more efficient was the scrubbing.

Phosphine is said to react similarly to AsH_3 with $KMnO_4$ to form Phosphonium Halides;

$$PH_3 + 2KMnO_4 = K_2HPO_4 + Mn_2O_3 + H_2O \quad (9)$$

it also reacts with NaOH to form:

$$PH_3 + 2NaOH + 2H_2O = Na_2HPO_4 + 4H_2 \quad (10)$$

Tests were performed with each of these media resulting in the graphs in Tables IV through VII.

TABLE IV: INLET CONCENTRATION OF EFFICIENCY TEST FOR PH_3 USING $KMnO_4$

FILE 25 RUN 3 STARTED 13:11.2 86/01/30 PH3/KMNO4 TEST XS METHOD 1
PH3/KMNO4 LAST EDITED 13:00.8 86/01/30 SMP AMT 1.0

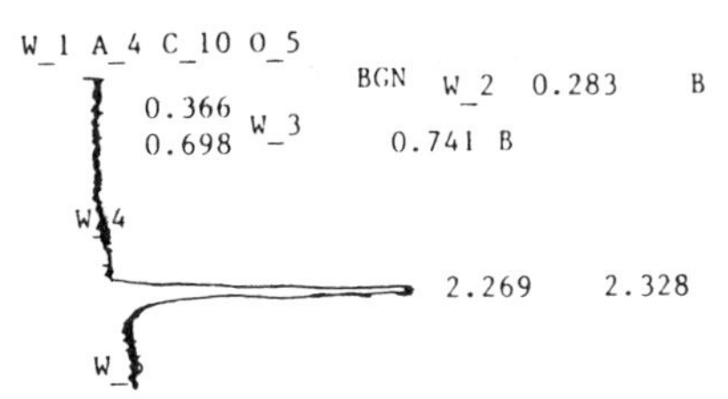

FILE 25 RUN 3 STARTED 13:11.2 86/01/30 PH3/KMNO4 TEST XS METHOD 1
PH3/KMNO4 LAST EDITED 13:00.8 86/01/30 SMP AMT 1.0

RT	AREA	BC	RT/10	RF	PPM	NAME
			0.274	3.151926E-03		PH3

2 MATCHED COMPONENTS 99.54% OF TOTAL AREA
5 UNKNOWN PEAKS UNRET PK 0.46% OF TOTAL AREA
7 PEAKS AREA REJECT 65066 TOTAL AREA

TABLE V: OUTLET CONCENTRATION OF EFFICIENCY TEST FOR PH_3 USING $KMnO_4$

FILE 27 RUN 5 STARTED 13:19.1 86/01/30 PH3/KMNO4 TEST XS METHOD 1
PH3/KMNO4 LAST EDITED 13:00.8 86.01/30 SMP AMT 1.0

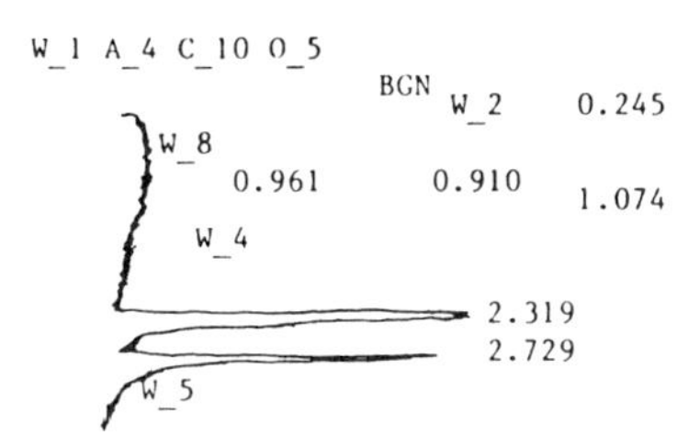

FILE 27 RUN 5 STARTED 13:19.1 86/01/30 PH3/KMNO4 TEST XS METHOD 1
PH3/KMNO4 LAST EDITED 13:00.8 86/01/30 SMP AMT 1.0

RT	AREA	BC	RT/10	RF	PPM	NAME
2.729	45940		0.273	3.151926E-03	144.7995	PH3

2 MATCHED COMPONENTS 99.67% OF TOTAL AREA
4 UNKNOWN PEAKS UNRET PK 0.33% OF TOTAL AREA
6 PEAKS AREA REJECT 120743 TOTAL AREA

arsenide:

$$2AsH_3 + (3FeCl_2) = Fe_3As_2 + (6HCL) \quad [9] \qquad (8)$$

A paper was given by E. Luthardt and H. Jurgensen of Aixtron [10] Gmbh in West Germany discussing the use of iodic acid to scrub AsH_3 with some success, we have concentrated our testing to the media ferric chloride, potassium permanganate, and potassium hydroxide. The ferric chloride results were not impressive, having only about a 45% efficiency. We then tested both $KMnO_4$ and a solution of NaOCL and KOH under the same conditions using a model SD 101 R effluent gas scrubber as the scrubber, and a Perkin Elmer Sigma 2000 Gas Chromatagraph with Photoionization Detector for data collection; Tables II and III show the results. Each test was performed a minimum of three times to verify the efficiencies.

TABLE II: LIQUID SCRUBBER EFFICIENCIES OF AsH_3 USING A $KMnO_4$ SOLUTION

TEST # 1

Test Gas: AsH_3
Concentration: 1.5% in N_2(15000000ppb)
Total Flow: 10 SLM
Test Liquid: 5%$KMnO_4$ in H_2O by weight
Liquid Volume: 25 gal.
Liquid Flow: 2.25 gpm
Test Equipment: Perkin-Elmer Sigma 2000 Gas Chromatagraph, LCI 100 Integration System
Calibration: P.I.D.-25000ppb AsH_3 in N_2

Test System: After scrubbing, the outlet of the scrubber was tested with a Perkin-Elmer Sigma 2000 Gas Chromatagraph first utilizing a hot wire detection sytem and after no readout was found, the P.I.D. system was used.

GRAPH FROM P.I.D. SYSTEM

File 49 Run 1 Started 11:44:1 35/09/10 AsH_3 US $KMnO_4$ 5% as method 3
AsH_3 EFF in FPD Last edited 11:43:4 85/89/18 SMP AMT 1.0

W_1 A_22 C_10 0_5
BGN AZ_ON W_2
W_3
W_4
AZ_ON
2.674 2.882 W_5AZ_ON

File 49 Run 1 Started 11:44.1 95/09/10 AsH_3 US $KMnO_4$ 5% XS Method 3
AsH_3 EFF in PPB Last Edited 11:43.4 85/89/19 SMP AMT 1.0

RT	AREA	BC	RT/10	RF	PPB	NAME
2.674	71531	T	0.267	4.914499E+00	344386.7500	Nitrogen
2.302	59020		0.280	4.314489E+00	284149.6250	Nitrogen
			0.362	2.889653E-01		Arsine

2 Matched Components 100.00% of Total Area
2 Peaks Area Reject 130551 Total Area

0.00 ppb Outlet reading equates to 100.0% Efficient

In the past, dilution of SiH_4 was sometimes used to reduce the possibility of explosion. This dilution was always done prior to the scrubbing system. As shown in Tables VIII and IX this dilution has a very adverse effect on the efficiencies of the scrubbing system. Another problem observed was that of a high negative draw on the scrubber : e.g., 2in. H_2O or greater. This high negative draw will not allow the SiH_4 gas to remain in contact with the scrubbing medium for a sufficient period of time for complete reaction. Therefore, limits must be put on dilution if proper scrubbing is to be achieved.

INSTALLATION AND MAINTENANCE

To ensure health and safety, it is extremely important to use common sense and to follow the manufacturers' recommended installation and maintenance procedures. All too often, we let the gas treatment system remain out of sight and out of mind. A plugging problem in a system can cause disastrous results, so proper maintenance is imperative. The gas treatment system must be treated with the same care that you would give any chemical reaction system. Purging the treatment system before and after cleaning is just as important a step as it is in the CVD system.

CATASTROPHIC RELEASE

The accidental release of a highly toxic, corrosive and/or flammable gas can cause severe injury to life and property. We have done extensive research on an emergency tank loss system for AsH_3. The results were an impressive 98% efficiency. Impressive is the fact that 98% could be achieved but even efficiencies of this magnitude do not reduce the AsH_3 concentration to below TLV or IDHL, therefore still posing a hazard.

Through all of our research, it became evident that the following parameters must be met:

1. The system must be on-line 100% of the time with no deterioration of the reaction medium.
2. The reaction medium must be able to remove the gas with close to 100% efficiency rate to be effective.
3. The system must be capable of handling multiple bottles and gases in order to be feasible.

Additional research is continuing, but no new results are available.

CONCLUSION

Given proper chemical solutions and maintenance of the scrubbing equipment, wet chemical scrubbing is a safe and viable way to effectively treat gaseous effluents from the photovoltaic industry. The scrubbing of AsH_3, PH_3, and SiH_4 can be done by utilizing simple systems in a single processing step. Any system used must be used in conjunction with the equipment manufacturer's recommendations.

REFERENCES

1. W. Braker, A. Mossman, Matheson Gas Data Book sixth edition,p291,1980.
2. Handbook of Chemistry and Physics, 63rd. ed., CRC Press, B-84, 1983.
3. Chem. Safety Data Guide, Bureau of Nat. Affairs, 1985.
4. Handbook of Toxic and Haz. Chem., M. Sittig, Noyes Publications, N.J., 1981.
5. Chemistry Made Easy, Snell & Snell, D. Van Norstrand Co., Vol.2, p.171, 1943.
6. Eng. Chem. Reactions Ref. Z Anal. Chem., 70, p105, 1908.
7. Gazz. Chem., D. Tivoli, Ital.,19, p631, 1889.
8. Z.Anorg. Chem.H. Reckleben, T. Scheiber, 70, p.255, 1911.
9. Enc. Chem. Reactions. ref., Z. Anorg. Chem.Brokl, 131, p.236,1923.
10. Hazmacon Proc. 86, E. Luthardt,H. Jurgensen, A.B.A.G., p.419, April 1986.
11. Handbook of Chem. and Phy., 63rd.ed., CRC Press, B143, 1983.
12. Matheson Gas Data Book,W. Braker, A. Mossman, 6th ed. , p.633,1980.

DESIGN AND PERFORMANCE CONSIDERATIONS FOR THE DEVELOPMENT OF HYDROGEN SELENIDE EFFLUENT PROCESSING SYSTEMS

William R. Bottenberg
Ashok Khanna
ARCO Solar, Inc., Chatsworth, CA 91311

Robert D. Sproull
Oregon State University, Corvallis, OR 97331

ABSTRACT

One of the gases which may be used in processes for fabrication of electro-optical materials including some solar cells is hydrogen selenide. This gas is included in the Environmental Protection Agency (EPA) list of toxic chemicals and requires care in use and treatment. No data were identified in the literature relating to scrubbing or other processes which have been used to control process effluents. This paper presents and discusses criteria for the development of an effective scrubbing system to handle hydrogen selenide based on current emission standards. Problems considered include the development of reactor and piping enclosure scrubbers as well as scrubbers for the effluent gas stream. Extrapolation of available data was used to predict performance of scrubbers based on typical designs. Data for the operation of a dual scrubbing system are presented which show that emissions of hydrogen selenide can be reduced well below threshold limit value (TLV)[1] for a range of gas throughput rates. Future areas of research for improving cost effectiveness and capacity are discussed.

Warning: Information in this report is furnished without warranty expressed or implied. No use of this information should be made without first obtaining the assistance of scientists, engineers, and professionals who have specialized training and experience in the use and handling of hazardous and toxic materials. The user of this information assumes any and all responsibility for its own use of this information and expressly relieves ARCO Solar, Inc., its parent, subsidiary and affiliated companies, Oregon State University and their respective officers, directors, agents and employees of any and all responsibility or liability for injury to property or persons including death and any loss, damage, or expense arising out of or in any way connected with such information.

INTRODUCTION

Modern processes for fabrication of semiconductor devices often require the handling of hazardous gases and subsequent neutralization or conversion to benign form. One process for the fabrication of copper indium diselenide solar cells requires the use of hydrogen selenide, H_2Se, gas.[2] This gas is classed as a hazardous chemical because of its toxic properties.[3,4] The threshold limit value (TLV) for the gas is very low: 50 ppb. Design of effluent processing systems is difficult since there are little chemical engineering data available for H_2Se. However, this gas is similar to a class of gases referred to as acid gases commonly found in the chemical processing industry. One of the commercially available and cost effective methods for removing acid gases such as CO_2 and H_2S from gas process streams is the use of a countercurrent packed column. In this method, a gas containing the component to be removed flows up the column and contacts liquid trickling down through the column packing, which provides a large interfacial area between the liquid and the gas. An effluent reactor of this design is often called a scrubber. Most acid gas scrubbers use a caustic liquid to convert the dissolved gas into an ionic form which has a very low vapor pressure.

PERFORMANCE MODELING OF SCRUBBERS

There are several important criteria for estimating the performance of scrubbers in processing effluent gases. The first basis depends on the equilibrium properties of the gas-liquid system. For instance, in scrubbing H_2Se with an aqueous KOH solution at high pH values, the following reaction is assumed:

$$H_2Se_{(g)} + 2\ KOH \iff Se^{=} + 2\ K^{+} + 2\ H_2O \qquad (1)$$

At the pH values used in the scrubbers, this reaction proceeds far to the right. Knowledge of the gas solubility and acidic ionization coefficients can be used to predict the resultant equilibrium partial pressure of the H_2Se in the gas stream as a function of the hydrogen ion concentration and total selenium content. Since these values change during the operation of typical scrubbers, it is important to be able to estimate the H_2Se partial pressure under a variety of conditions. The overall equation (1) for the reaction actually consists of several intermediate steps as shown below:

$$H_2Se_{(g)} \iff H_2Se_{(d)} \qquad (2)$$

$$H_2Se_{(d)} + OH^{-} \iff H_2O + HSe^{-}\ \text{(biselenide ion)} \qquad (3)$$

$$HSe^{-} + OH^{-} \iff H_2O + Se^{=}\ \text{(selenide ion)} \qquad (4)$$

Equation (2) represents the dissolution of the gas into the liquid to form a dissolved species while equations (3) and (4) show successive ionization of the dissolved species.

An equally important performance consideration is the rate of reaction in equations (2)-(4) as a function of the gas flow rate, liquid flow rate, gas and liquid composition, and temperature. Equilibrium data may only be used to estimate performance of a system at very low throughput rates while rate data should generally be used to estimate performance under actual operating conditions. An excellent set of references for chemical engineering calculations on packed column scrubbers can be found in Treybal[5] and Perry[6]. An adequate description of the performance of packed column scrubbers in the range of conditions expected to be found in typical semiconductor applications (i.e., dilute gas mixture and instantaneous reaction in the liquid) is given by the following equation:

$$\ln(y_1/y_2) = G / (K_G a\ P\ Z) \quad (5)$$

where y_1 and y_2 are the entering and exiting mole fractions of the gas component to be scrubbed, G is the total gas flow rate per cross-sectional area of column, P the total pressure of the gas, Z the height of the column packing, K_G the overall gas-phase mass-transfer coefficient, and a the interfacial surface area per unit volume of packing. The use of this equation implies that the composition of the liquid as it flows past the gas in the column does not significantly change from top to bottom of the column. Thus, the objective of an analysis of system performance is to calculate the composition of the gas phase at the column exit as a function of gas flow and overall gas-phase mass-transfer coefficient.

PHYSICAL DATA FOR H_2Se

H_2Se is a gas in the same family as hydrogen sulfide, H_2S. It is slightly soluble in water and forms an acidic solution since it acts as a weak diprotic acid. The physical chemistry data for H_2Se are limited, consisting of Henry's law values and ionization constants.[3,7] These data are presented in Table I along with data for H_2S and CO_2, which are also classed as acid gases.

Table I. Physical chemistry data for H_2Se, H_2S and CO_2.

	H_2Se	H_2S	CO_2
Henry's Law Coef.			
H, mol/l·atm	0.33	0.099	0.033
Ionization Const.			
K_1, mol/l	1.3×10^{-4}	9.1×10^{-8}	4.2×10^{-7}
K_2, mol/l	1.0×10^{-11}	1.1×10^{-11}	4.8×10^{-12}

The Henry's law constant corresponds to values of total H_2Se concentration near infinite dilution. The activity coefficients for the solution components are assumed to be 1. As will be discussed later, there are no data for the important H_2Se solubility reduction in concentrated ionic solutions.

The parameter required for estimation of scrubber performance is the value of K_Ga over the operating range of the scrubber.

SCRUBBER DESIGN

The scrubbers described in this paper actually consist of several components: an ejector venturi input section, a holding tank for the scrubbing fluid, and a packed column output section. As shown in Fig. 1, the effluent gas enters the scrubber into an ejector venturi unit. This section uses a pump and nozzle to spray the scrubbing solution at high velocities through a narrowed pipe section past the gas entry port. This high velocity spray entrains the gas in a turbulent stream permitting reaction with the gas and removal of particulates. The high velocity stream also creates a partial vacuum at the input port which tends to cause the input line to operate at a lower pressure with respect to the reactor output port. The sprayed liquid is recirculated through a filtered pump to remove particulate matter which can form in the scrubber due to the oxidation of selenide ion to selenium. The gas then moves to the packed column output section. The gas enters the column from below and rises to the top de-misting section. The liquid medium is pumped to the top of the column and sprayed over a bed of plastic packing material. As it descends, it contacts the rising gas and absorbs the reactive gas component. At the bottom of the column, it drains into the common liquid storage chamber ready for recirculation to the top of the column. Provision is made for the addition of water over time since the flow of the dry gas through the system causes significant water evaporation.

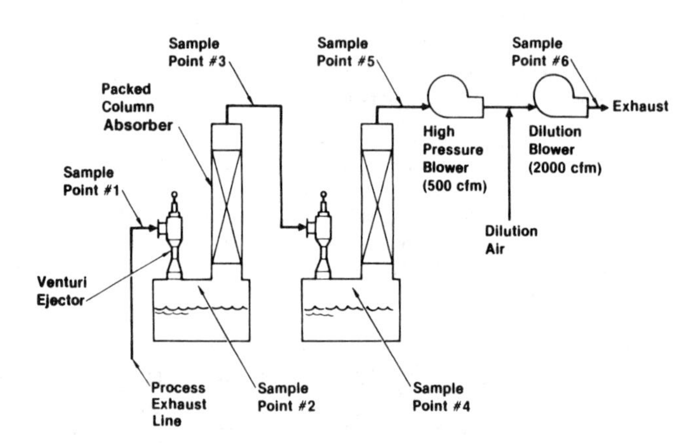

Fig. 1. Diagram of scrubber for H_2Se process gas treatment.

During a typical cycle of operation, the process gas effluent scrubber is charged with 25-30 gallons of 7% KOH and operated until the pH lowers to a threshold value. The system is then drained and recharged. During the course of operation, the selenide ion is typically converted to solid selenium by air oxidation according to the following reaction:

$$2\ Se^{=} + O_2 + 2\ H_2O <=> 2\ Se_{gray} + 4\ OH^{-} \tag{6}$$

The source of the oxygen is from the pump/purge cycling of the reactor. The gray form of selenium is the thermodynamically stable form in aqueous solution. As a result of this process, the system continues to partially regenerate forming the reactive hydroxide ion by air oxidation. The life of the solution is actually limited by the tendency of the solution to absorb CO_2 from air, particularly at high pH conditions:

$$CO_2 + 2\ OH^{-} <=> CO_3^{=} + H_2O \tag{7}$$

During normal operation, air does not flow through the reactor and effluent scrubber system, however during pump/purge portions of the process cycle, air will flow through the system.

SCRUBBER OPERATION PERFORMANCE RANGE

The range of operation performance for the scrubber can be bracketed by calculating the dependence of the equilibrium vapor pressure of H_2Se on the pH of the solution and on the total selenium content of the solution during the course of the operation of the scrubber with an initial charge of caustic. Figure 2 shows curves of constant H_2Se vapor pressure as a function of solution pH and the logarithm of the total selenium content of an equilibrium solution in moles per liter. The calculations were performed using Table I values and the equilibrium equations (2)-(4). The curves correspond to vapor pressures of 1 ppb, 50 ppb (TLV value), 1 ppm, and 2 ppm. There is no universal standard for the concentration of a hazardous material in an effluent gas stream, but it certainly should be below the TLV value for areas where people may be found to assure that no unhealthful exposure will occur. H_2Se is a gas with a disagreeable smell apparently detectable by people in the range of 0.5 to 2 ppb, values well below the TLV. However, smell is not a reliable monitoring or detection method and olfactory fatigue can quench any noticeable odor. Appropriate hydride sensors are generally applied to applications requiring detection.[8] Figure 3 shows the decrease in pH as a function of the fraction of the initial caustic charge which has reacted. Figure 3 suggests that the overpressure of H_2Se may be controlled to levels below the level of detectability by implementing an approach to maintain pH above 11 while preserving adequate aeration. The conditions assume that the caustic started as a 1 g-mol/liter solution of KOH. The rapid decrease in pH at fraction 0.5 can be used as an indicator that the solution is ready for change. Prudent operation of the scrubber would require a fail-safe source interlock condition for pH lower than 11.0. It is important to recall that the pH will decrease also as a result of the CO_2 reaction shown in equation (7). Figure 4 shows the dependence of the logarithm of the H_2Se overpressure as a function of the fraction reacted. The value of -9.0 corresponds to 1 ppb. This analysis is straightforward enough, but it does not account for the fact that the solubility of H_2Se is very likely to be lower in a solution of high ionic strength. Since no data are available for H_2Se solubility as a function of ionic strength, careful measurements of scrubber performance are required to assure that the operating boundaries are appropriate. To assure that the systems used in initial design qualification testing would be safe, identical scrubbers were linked in series. Each scrubber was calculated to be capable of independently handling H_2Se at operating flow rates.

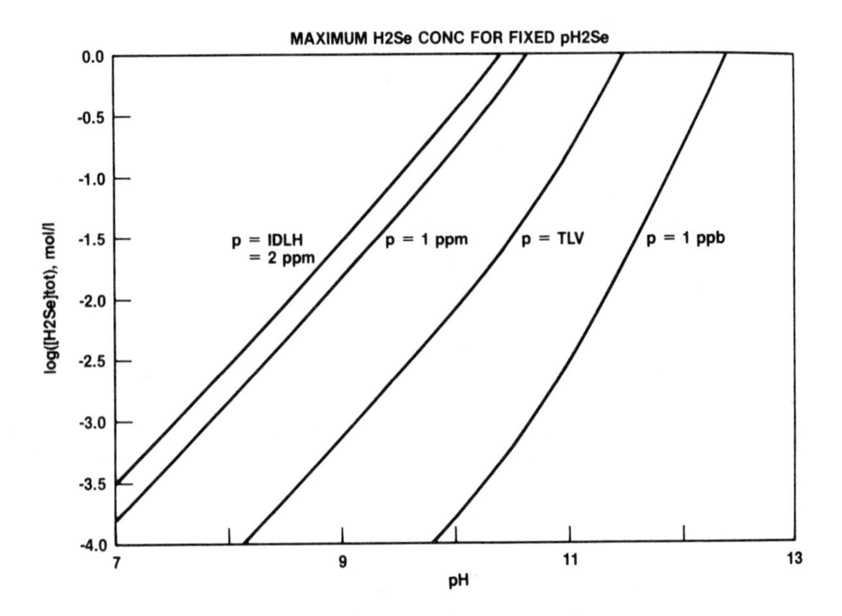

Fig. 2. Equilibrium H_2Se vapor pressure as a function of pH and total selenide content of solution.

The efficiency of the scrubber was calculated as a function of flow rate using equation (5) for several values of K_Ga: 5, 7.5, and 10 lb-moles/hr-ft^3-atm. These values spanned the range of estimated values for this parameter for H_2Se based on similarity to H_2S and CO_2 values in comparable reactors.[5,6] The results of these calculations are shown in Fig. 5. The maximum flow rate for the system was specified to be 15 scfm based on the requirement that acceptable efficiencies for the scrubbers be obtained for the range of probable K_Ga values. The operating pressure is 1 atmosphere. The efficiency is defined as $1-y_2/y_1$.

PERFORMANCE MEASUREMENTS

The performance measurements for the scrubber system were conducted while flowing various concentrations of H_2Se and nitrogen through the scrubber at different flow rates while measuring the H_2Se concentration at several points in the system. The concentration was measured using a Photovac Model 10S70 portable gas chromatograph calibrated with a standard H_2Se sample. The points of measurement were: scrubber inlet, solution chamber space between venturi ejector and packed column inlet, scrubber outlet, and dilution exhaust outlet. No H_2Se was detected in the scrubber for the concentration and flow conditions evaluated. Table II displays these conditions. An important outcome of the trials was that the venturi ejector was demonstrated to be highly efficient. The detection limit of the GC was calculated to be 5 ppb based on baseline measurements.

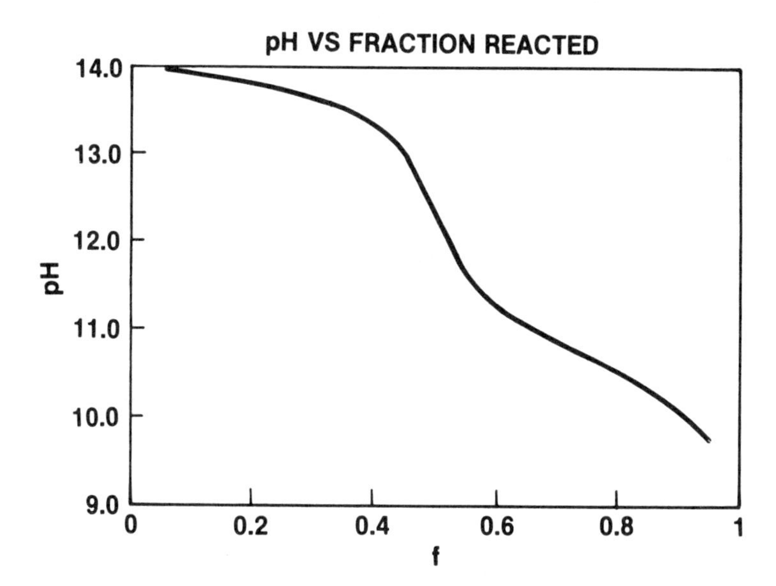

Fig. 3. pH as a function of fraction of solution reacted.

CONCLUSIONS AND OUTLINE FOR FUTURE WORK

Dual scrubbers provide several important features, especially for the exploratory development discussed in this review. Each scrubber was capable of handling the normal gas process stream, so that if one failed, the system would perform correctly. The dual system was also capable of controlling a gas cylinder upset condition such as valve failure. This was designed into the system by the use of a flow limiting restrictor at the cylinder head.

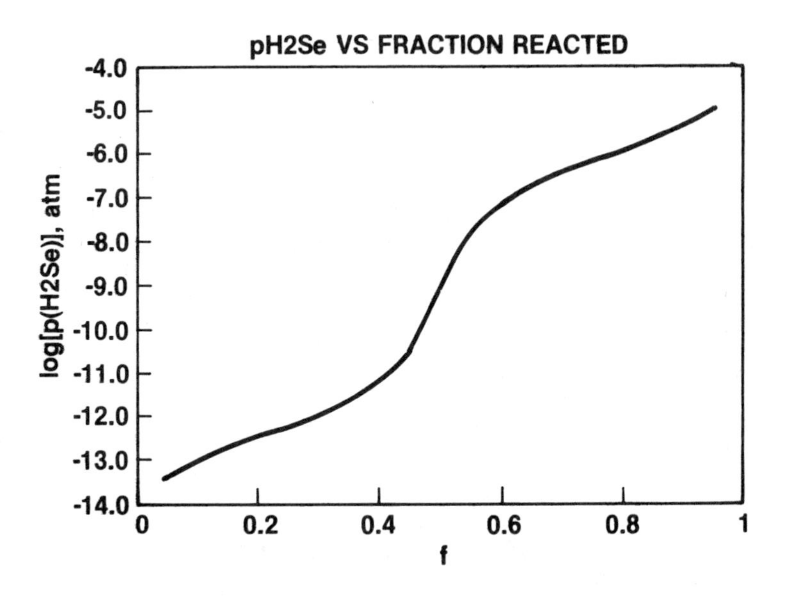

Fig. 4. Vapor pressure of H_2Se as a function of fraction of solution reacted.

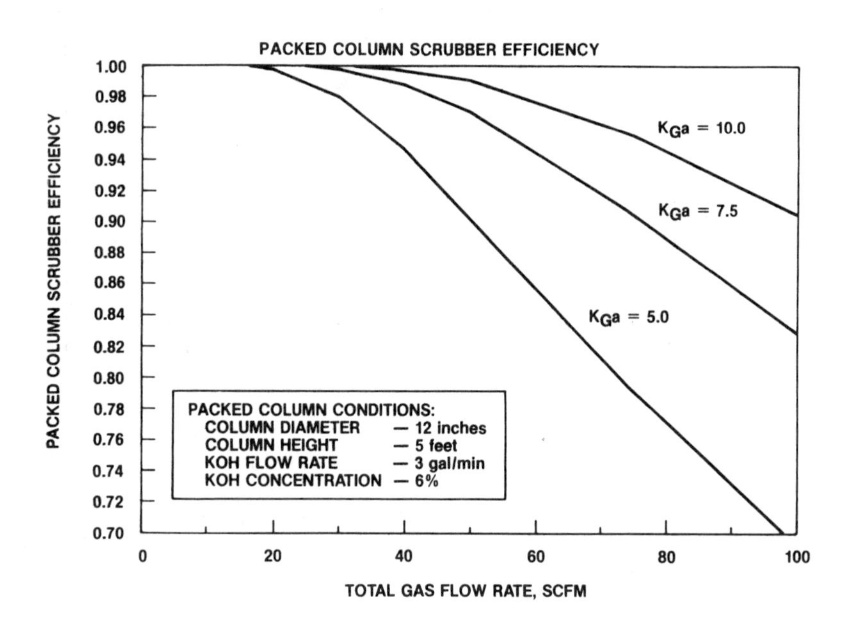

Fig. 5. Calculated scrubber efficiency vs flow rate for K_Ga values ranging from 5 to 10 lb-moles/hr-ft^3-atm.

Table II. Scrubber performance testing conditions.

H_2Se Flow Rate sccm	N_2 Flow Rate scfm	H_2Se Concentration Inlet (calc)	Chamber ppb (meas)	Outlet ppb (meas)	Exhaust ppb (meas)
10.0	1.0	350 ppm	ND	ND	ND
50.0	20.0	87 ppm	ND	ND	ND
200.0	22.0	0.012%	ND	ND	ND
3228.0	52.0	0.19%	ND	ND	ND

ND=Not Detected.

There remain several important areas for future work. The systems as designed are acceptable for research purposes, but improved economy is possible for commercial application. The actual values of K_Ga over the operating range can be further refined. These values will permit more accurate prediction of the performance of commercial systems as a function of size and throughput. In addition, the equilibrium values for the vapor pressure of H_2Se as a function of operating conditions can be refined so that the operating boundaries which define safe and odor-free operation of scaled systems can be calculated.

The presence of CO_2 in the process stream will reduce the operating time for a given charge of caustic since it irreversibly reacts with the hydroxide under the expected operating conditions for the scrubber. It is inevitable that some CO_2 will be present because of the necessity for air pump/purge cycles in the process sequence. Approaches toward minimizing the effect of CO_2 on the operating performance of the scrubber while optimizing the collection of solid by-product selenium for recycling would be useful.

It is clear that safe, reliable scrubbers for H_2Se can be designed. Work remains to be done to optimize cost-effective designs for large-scale operation.

REFERENCES

1. The time-weighted Threshold Limit Value (TLV) is the time-weighted average concentration for a normal 8-hour workday and a 40-hour workweek, to which nearly all workers may be repeatedly exposed, day after day, without adverse effect.

2. U.S. Patent 4,581,108, Process of Forming a Compound Semiconductive material.

3. The Merck Index (Merck & Co., Rahway, N.J. 1976).

4. N.I. Sax, and R.J. Lewis, Rapid Guide to Hazardous Chemicals in the Workplace (Van Nostrand Reinhold Co., New York, N.Y. 1985).

5. R.E. Treybal, Mass-Transfer Operations, 3rd ed. (McGraw-Hill, New York, N.Y. 1980).

6. R.H. Perry and C.H. Chilton, Chemical Engineers' Handbook, 5th ed. (McGraw-Hill, New York, N.Y. 1973).

7. R.C. Weast (ed.), Handbook of Chemistry and Physics, 57th ed. (CRC Press, Cleveland, OH. 1977).

8. J.J. Roy, Solar Cells 19, 263 (1986-87).

CONTROLLED COMBUSTION, DECOMPOSITION AND OXIDATION -- A TREATMENT METHOD FOR GASEOUS PROCESS EFFLUENTS

Roger J. B. McKinley, Sr.
Innovative Engineering, Inc., Santa Clara, CA 95054

ABSTRACT

The safe disposal of effluent gases produced by the photovoltaic industry deserves special attention. Due to the hazardous nature of many of the materials used, it is essential to control and treat the reactants and reactant by-products as they are exhausted from the process tool and prior to their release into the manufacturing facility's exhaust system and the atmosphere.

Controlled combustion, decomposition, and oxidation (CDO) is one method of treating effluent gases from photovoltaic thin film deposition processes. This report discusses CDO equipment applications, field experience, and results of the use of CDO equipment and technological advances gained from field experiences.

INTRODUCTION

A number of extremely hazardous gases are routinely used by the semiconductor and photovoltaic industries to manufacture integrated circuits and other devices. Although the processes generally consume only small quantities of these gases, the nature of the gases is such that even small amounts of gas can do considerable damage to people and equipment in a very short time.

Effective controlled process exhaust gas conditioning has been and is becoming an issue of growing concern for specialty gas processors, process equipment manufacturers, users of process equipment various governmental agencies, inusrance underwriters, and the general public.

CODES AND REGULATIONS

During the last few years, there has been and continues to be growing concern over finding safer methods of storing, dispensing, and monitoring these gases. More recently the disposal of these gases has been receiving greater and greater attention.

In response to growing concern over toxic substance problems, Congress has enacted over two dozen regulatory statutes covering the routes by which certain chemicals or aspects of chemical use can threaten human health and the environment.

The laws are administrated by various federal regulatory agencies. The Occupational Safety and Health Administration (OSHA) has been given responsibility to insure a safe workplace. It sets mandatory job safety and health standards and provides reporting procedures for all industrial injuries and fatalities. The Environmental Protection Agency (EPA) attempts to prevent further poisoning of the environment by requiring industry to develop air quality control. The Federal Water Pollution Control Act dictates controls to improve water quality. The Toxic Substances Control Act regulates the use of certain toxic materials and establishes a toxic substance data base, while the Resource Conservation and Recovery Act gives EPA added authority over how waste materials are disposed.

The Department of Transportation sets standards for labeling (DOT Hazard System), packaging, testing, and handling compressed gases. Other agencies such as NIOSH (part of OSHA), the Uniform Building Code (UBC) and its companion document, the Uniform Fire Code (UFC), and other codes and state and local regulations have meaningful mandates that also deserve our attention.

The UBC published by the International Conference of Building Officials is the most widely used model building code in the country. It is the code for the western United States. The Basic/National Building Code is used primarily in the northeastern part of the country, and the Standard Building Code is used almost exclusively in the southeastern part of the country.

Occupancy is one of the primary regulatory criteria in the building code and is based on the use or occupancy in the proposed building. The seven major classes or Occupancy Groups are:

A - Assembly
B - Business
E - Educational
H - Hazardous
I - Institutional
M - Miscellaneous
R - Residential

Within each of these Occupancy Groups are sub-categories called Divisions. In Group H Occupancy, there are six Divisions; and Division 6 may be briefly defined as semiconductor fabrication facilities.(2)

"Division 6. Semiconductor fabrication facilities and comparable research and development areas when the facilities in which hazardous production materials are used and constructed in accordance with Section 911 and when storage, handling, and use of hazardous materials is in accordance with the Fire Code." (2)

The UFC, which is the product of the Western Fire Chiefs Association, is a companion document to the UBC. Many provisions of the codes are interrelated and cross referenced. Administratively, the fire code is organized into eight basic parts consisting of one or more articles in each part. (4)

Part I Administration
Part II Definitions and Abbreviations
Part III General Provisions for Fire Safety
Part IV Special Occupancy Use

Part V Special Processes
Part VI Special Equipment
Part VII Special Subjects
Part VIII Appendices

Article 51 of the UFC, which is companion to H-6 of the UBC, appears in part V of the fire code. Article 51 deals with controls required for the utilization of hazardous materials in the production of semiconductor devices and related research functions. The required controls relate to the nature of materials encountered, their physical state, and the condition in which they are found in the building, i.e., storage, handling, or use. Control over the materials is achieved by limiting the amount of material in use at any particular work station and applying engineering controls such as sprinklers, automatic leak detection, local exhaust, and warning systems. (4)

The purpose of the UBC and UFC is to protect the public by regulating the construction, alteration, and maintenance of structures and the storage, handling, and use of hazardous materials. The codes establish certain minimum criteria that define the code intent. The UBC and UFC are concerned with fire and related hazards. The codes may be adopted locally or on a statewide basis. The adoption may be made by ordinance or by a legislative act.

An excellent reference concerning H-6 Occupancy is available through Larry Fluer, Inc., P.O. Box 10386, San Jose, CA 95157.

Two new pieces of "legislation" confronting the semiconductor wafer fabrication industry and other institutions using hazardous materials are:

1. Uniform Fire Code - Article 80 Rewrite
2. California Assembly Bill #1021

Previously the codes primarily addressed the storage, dispensing, and monitoring of hazardous materials. With the addition of Article 80, the codes now address STORAGE, DISPENSING, MONITORING, and DISPOSAL.

CONTROLLED PROCESS EXHAUST GAS CONDITIONING

Effective controlled process exhaust gas conditioning has been and is becoming an issue of growing concern for specialty gas processors, process equipment manufacturers, users of process equipment, various governmental agencies, insurance underwriters, and the general public.

Exhaust gas conditioning or removing harmful substances from process exhaust gases is not necessarily needed for succesful device wafer fabrication; and perhaps since exhaust gas conditioning equipment is not revenue producing, its importance has not been previously fully acknowledged. However, Factory Mutual, a leading insurance company for semiconductor manufacturers, reports $60,000,000 in claims were paid for 112 incidents from 1974 through 1986. The $60,000,000 reported by Factory Mutual does not include claims covered by other underwriters, unreported losses, and self-insured claims. Fire was by far the most frequent cause of loss and the ignition of flammable and pyrophoric gases was the leading cause of fire. Hydrogen and silane gases were involved in approximately 90 percent of the reported incidents.

Table I: SEMICONDUCTOR PLANT LOSSES
1974 THROUGH 1986 $60,000,000

CAUSE	NUMBER OF INCIDENTS
Fire	72
Liquid Leakage or Spillage	29
Human Element	11
TOTAL	112

Table II: CAUSES OF FIRES AT SEMICONDUCTOR PLANTS

CAUSE	NUMBER OF INCIDENTS
Ignition of Flammable or Pyrophoric Gases (27 of the 29 or approximately 90% involved Hydrogen or Silane)	19
Electrical Origin	20
Immersion Heaters	13
Ignition of Flammable Liquids	6
Hot Plates	4
TOTAL	72

Although exhaust gas conditioning may not be essential for semiconductor wafer fabrication, it is important for the protection of personnel, protection of the environment, and, as stated above, the protection of the manufacturing facility.

The contaminants encountered in process exhasut gas streams are extremely varied. Rarely does an exhaust stream contain only one classification of contaminant -- whether it be particulate, gas, or vapor. While the gases introduced into almost all semiconductor wafer fabrication processing chambers are well defined, the composition of the gas mixture exiting the process chamber is generally not precisely defined. Furthermore, since the mid 1970's, there has been an ever increasing utilization of subatmospheric or reduced pressure (vacuum) processing in semiconductor wafer fabrication. This has introduced additional unknowns such as pump oil vapors into the process effluent.

For the most part, reactive gases are not released unreacted from a process chamber; and in most cases, the gases are present only in low concentrations. However, almost all are highly toxic (AsH_3, PH_3, B_2H_6), pyrophoric (SiH_4), flammable (H_2), or corrosive (HCl). Hydrogen (H_2) is a by-product of all hydride (SiH_4, AsH_3, B_2H_6, PH_3) reactions; and under controlled exhaust conditions, SiH_4 can be the ignition source for a serious H_2 reaction.

"THE PENINSULA TIMES TRIBUNE

DAMAGE MAY RUN HIGH IN FIRE A LOCAL PLANT
MOUNTAIN VIEW -
...The fire, ignited by gas and vaporized oil, spread through an air duct system in a fabrication room where valuable silicon chips are manufactured.
...The fire began in an air duct system where vaporized oil from a vacuum pump and gas from a malfunctioning reactor had collected."

No manufacturer wants a mishap due to improperly handled exhaust gases to end up in the morning paper.

In the past, dilution or simple water washing was employed to dispose of exhaust gases. Dilution to "safe" levels has long been an accepted practice, but public sentiment and new codes are diametrically opposed to any practice in which chemicals of any kind are released into the atmosphere untreated.

Conscientiously, we should address pounds per hour, not parts per million (ppm).

"DILUTION IS NO SOLUTION TO POLLUTION!"

Plain water scrubbing is ineffective for most of the hazardous gases encountered in wafer fabrication.

SOURCE VS. CENTRAL EXHAUST GAS CONDITIONING

Exhaust gas conditioning equipment may be divided into two broad categories -- source and central systems.

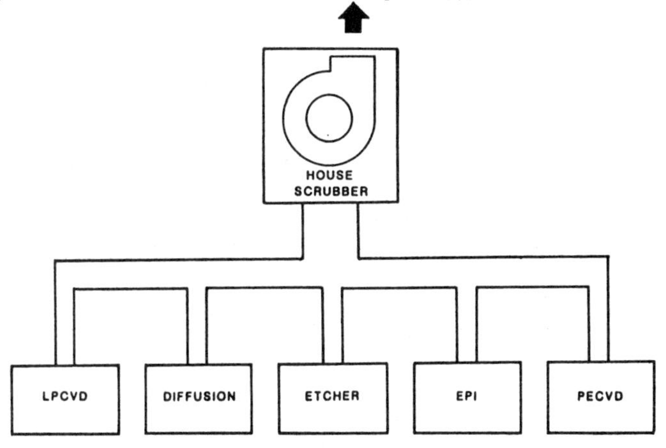

Fig. 1. Central Conditioning - Multi process tools with incompatible exhaust effluent entering common duct system and being transported to remotely located exhaust treatment system.

Because of the nature and low volume of effluent exiting most wafer fabrication systems, it is desirable and more effective to conditioning process exhaust gases as close to their source as is physically possible.

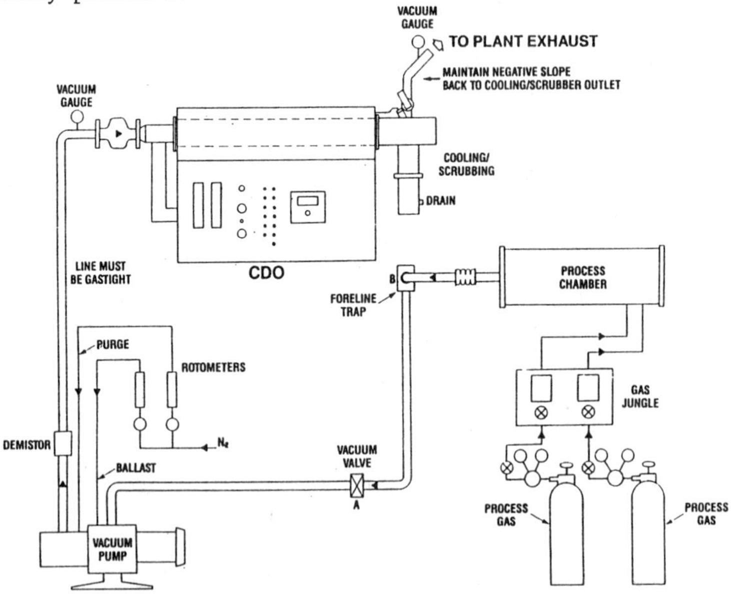

Fig. 2. Source Conditioning - Dedicated gas treatment system located as close to process tool exhaust as physically possible.

Because of the nature of the gases and vapors exiting most process systems and based on the interpretation of codes and regulations, source conditioning may be mandatory in the future. From UFC Article 51: "Duct Systems - Reactives. Two or more operations shall not be connected to the same exhaust system when either one or the combination of the substances may constitute a fire, explosion, or chemical reaction hazard within the duct system." (4)

Also, Section 1105 of the 1985 Edition of the Uniform Mechanical Code: "A mechanical ventilation or exhaust system shall be installed to control, capture, and remove emissions generated from product use or handling when required by the Building Code or Fire Code and when such emissions result in a hazard to life or property. The design of the system shall be such that the emissions are confined to the area in which they are generated by air currents,hoods, or enclosures, and shall be exhausted by a duct system to a safe location or treated by removing contaminants. Ducts conveying explosives or flammable vapors, fumes, or dusts shall extend directly to the exterior of the building without entering other spaces. Separate and distinct systems shall be provided for incompatible materials."

CONTROLLED "SOURCE" PROCESS EXHAUST GAS CONDITIONING IS:

1. THE WAVE OF THE FUTURE
2. EVOLUTIONARY/EVOLVING
3. MORE EFFECTIVE
4. DESIRABLE
5. SAFER
6. ESSENTIAL TO COMPLY WITH CODES

PROCESS TOOL EXHAUST

Many of the process systems in use today and proposed for the future are sub-atmospheric pressure systems utilizing vacuum pumps.

The required pumping systems for sub-atmospheric processes are fairly well defined and for simplicity may be broken down into three areas (see Fig.1): foreline (inlet), pump, and exhaust.

That portion of the pumping system connecting the process chamber to the vacuum pump may be called the foreline or inlet. The foreline generally includes a flexible stainless steel interconnect line, a particulate trap, and a vacuum valve. Pumping precautions which should be taken include:

(a) preventing the condensation or trapping of chemicals in the pump rotor and stator area;
(b) preventing oxygen from entering the pump (i.e., through the ballast valve);
(c) preventing the accumulation of explosive, toxic, and/or corrosive gases in the dead volume of the pump oil reservoir.

Items (a) and (b) may be accomplished by injecting a 1 to 2 SLM dry nitrogen flow into the pump ballast inlet. Since most pumps are equipped with an inlet to the pump oil chamber, item (c) may be accomplished by flowing 2 to 3 SLM of dry nitrogen into this inlet. Another effective technique for the oil chamber purge is to bubble the dry nitrogen through the pump oil. This dry nitrogen flow will help to dilute and eject the explosive, toxic and/or corrosive gases from the pump oil chamber. Ideally, regulators, flow meters, and pressure gauges should be used to set and monitor the dry nitrogen gas flow.

It should be noted that large flows of purge gas through the oil chamber may result in an unreasonable loss of fluid from the pump due to rapid removal of fluid vapor. Excessive nitrogen purge may also dilute the exhaust gases to such a diluted concentration as to be detrimental to effective exhaust gas conditioning.

The effluent exiting the pump casing frequently contains pump fluid vapors and toxic and flammable or explosive gases. An uncontrolled mixing of the reactive gases (i.e., silane or hydrogen) with air can create explosive conditions. Uncontrolled discharge of toxic gases (i.e., arsine, diborane, and phosphine) can create environmental as well as personnel hazards. Unconditioned vacuum pump fluid vapors can condense in exhaust ducting causing maintenance and health and safety hazards.

The toxicity and flammability of vacuum pump effluent is hazardous enough that particular attention should be given to "gas-tightness" of the entire high pressure (exhaust) section and to the "conditioning" or treatment of the pump effluent prior to its release into the atmosphere.

The design of the gas exhaust line mounted on the discharge port of a mechanical pump should follow certain basic reules (see Fig.2):

a) Exhaust lines should be sized so as not to create pump discharge back pressure.
b) The exhaust line should be constructed of gas tight metal tubing. Stove pipe type 'jointed' ducting should not be used as it is not air tight. Air must not be allowed to mix with the gases prior to conditioning.
c) The line should have slightly sub-atmospheric pressure (½ inch to 1 inch of water below atmosphere) to assist in ejecting the effluent. However, this pressure should not be so low (i.e., 8 to 9 negative inches of water) that excessive pump oil vapors are sucked from the pump.
d) A pressure gauge should be installed on or just downstream of the pump exhaust port to monitor exhaust line pressure.
e) The exhaust line should incorporate a device to condense fluid vapors from the pump effluent.
f) "Conditioning" of the pump effluent is necessary for personnel safety, protection of property, and protection of the environment.
g) Pressure (vacuum) gauges should be mounted on the inlet and discharge sides of the conditioning equipment.

While the gases introduced into the process chamber are generally well known, the composition of the gax mixture exiting the process chamber, passing through, and exiting the pump is generally not precisely defined. However, the pump effluent generally contains enough toxic, explosive, and combustible materials that effluent "conditioning" is essential

TABLE III: TYPICAL SUBATMOSPHERIC CVD GASS EFFLUENT

CVD PROCESS	PROCESS GASES	PROCESS EFFLUENT
Poly	SiH_4	Si, SiH_4, H_2, Pump Fluid Vapors
Doped Poly	SiH_4, PH_3	Si, SiH_4, PH_3, H_2 Pump Fluid Vapors
Low Temperature Oxide (400°C)	SiH_4, O_2, PH_3	SiO_2, SiH_4, O_2, PH_3, H_2, Pump Fluid Vapors
Tungsten?Tungsten Silicide	SiH_4, WF_6. H_2	WF_6. WFx, SiH_4, SiHx, H_2, N_2, Pump Fluid Vapors

PROCESS EXHAUST GAS CONDITIONING

Controlled combustion is a method of reducing the toxic and flammable hazards of vacuum pump effluent.

Process exhaust gases enter the Controlled Combustion, Decomposition, Oxidation (CDO™) System (see Fig. 3) through the unit's flame check and under controlled conditions are mixed with an oxygen source in the oxygenator. The oxygen enriched gases then flow through a high temperature section where combustion takes place and exit the CDO™ System through a water mist cooling and scrubber section. Tases and pump oil vapors and combusted and oxidized.

In the CDO™ unit, the pump effluent is oxygen enriched under controlled conditions and exposed to a high temperature environment thereby increasing the liklihood of a complete chemical reaction.

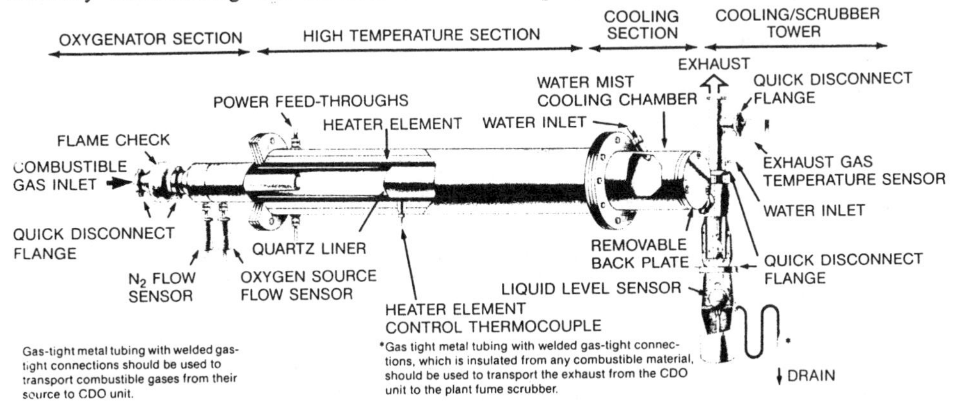

Fig 3. Controlled Combustion, Decomposition, and Oxidation System Cross Section

Control of time, temperature, turbulence, and oxygen is essential for complete reaction. Time is the period of residence of the waste gas(es) in the combustion chamber. Time is a function of combustion chamber length and diameter and gas volume.

Temperature is the temperature of the waste gas(es) in the combustion chamber. Turbulence is necessary to insure the proper mixing of the waste gas(es) and the oxygen source. Oxygen is required to support the complete reaction of the waste gas(es).

As reported by Arco Solar, Inc. in their January, 1987, "Evaluation of Hazards Associated With the Manufacture of Thin Film Solar Cells for Brookhaven National Laboratory": "The CDO approach has been shown to be highly effective in the controlled combustion of silane and diborane, the results suggesting performancy efficiency approaching 100% for silane. However, under the defined operating conditions, the system appears less efficient in the removal of phosphine." (11)

Since publication of the Arco report, Innovative Engineering Inc. has performed extensive testing with arsine, diborane, and phosphine.

REFERENCES

1. Henon, B.K., PhD., "Double Containment Gas Handling", Semiconcuctor International (April, 1987).
2. International Conference of Building Officials, "Uniform Building Code - 1985 Edition" (issued every three years), 5360 South Workman Mill Road, Whittier, CA 90601.
3. Goldberg, A. and Fluer, L., "H-6 Design Guide to the Uniform Codes for High Tech Facilities," (December, 1986).
4. International Conference of Building Officials and Western Fire Chiefs Association, "Uniform Fire Code - 1985 Edition," (issued every three years), 5360 S. Workman Mill Road, Whittier, CA 90601.
5. Pytrga, F., President, PAACO Inc., "Emergency Containment System," Semiconductor Safety Association (January, 1987).
6. Bolmen, R. A., Siliconix, Inc., "Hazardous Production Gases: Part 1: Storage and Control," Semiconductor International (April, 1986).
7. Burggraff, Peter, "Emphasizing Effluent Gas Scrubbing," Semiconductor International (June, 1984).
8. Librizzi, J. and Manna, R. R., Heat Systems Ultrasonics Inc., "Controlling Air Pollution from Semiconductor Fabrication Operations," Microelectronic Manufacturing and Testing (December, 1983).
9. Fluer, Larry, "Gases for Microelectronic Manufacturing: Storage, Handling and Use," Microelectronic Manufacturing and Testing, (November, 1983).
10. Herb, G. K., et al, "Plasma Processing: Some Safety, Health, and Engineering Considerations," Solid Atate Technology (August, 1983).
11. Khanna, Ashok K., et al, "Evaluation of Hazards Associated With the Manufacture of Thin Film Solar Cells," (January, 1987).
12. McKinley, R.J.B., "Vacuum Equipment Exhaust Considerations for Subatmospheric Pressure CVD Applications," Solid State Technology, (August, 1987).
13. Dillon, A.P., "Hazardous Waste Incineration Engineering."

SAFE GAS HANDLING AND SYSTEM DESIGN FOR THE LARGE SCALE PRODUCTION OF AMORPHOUS SILICON BASED SOLAR CELLS

C.M. Fortmann, M.V. Farley, M.A. Smoot and B.F. Fieselmann
Solarex Thin Film Division, 826 Newtown-Yardley Rd., Newtown, PA 18940

ABSTRACT

Solarex is one of the leaders in amorphous silicon based photovoltaic production and research. The large scale production environment presents unique safety concerns related to the quantity of dangerous materials as well as the number of personnel handling these materials. The safety measures explored by this work include gas detection systems, training, and failure resistant gas handling systems. Our experiences with flow restricting orifices in the CGA connections and the use of steel cylinders is reviewed. The hazards and efficiency of wet scrubbers for silane exhausts are examined. We have found it to be useful to provide the scrubber with temperature alarms.

INTRODUCTION

The solar cell industry, as well as the semiconductor industry in general, is dependent on a variety of toxic and dangerous materials which only recently have come into large scale use. The dangers involve materials that pose both a toxicity problem and a fire hazard. It is essential that toxic materials not be released into the work place or the surrounding environment. It is equally important that materials that readily ignite in air not be allowed to either accumulate or be uncontrollably released. In this presentation we will attempt to share our experience with the handling of the toxic and dangerous materials associated with the production of amorphous silicon solar cells.

Materials used in the production of amorphous silicon include silane (SiH_4), phosphine (PH_3), diborane (B_2H_6), and hydrogen (H_2). Recently, germane (GeH_4) has been finding wider use for the preparation of amorphous layers. Tin oxide coatings typically employ either $SnCl_4$ or tetramethyltin. A general review of the hazards associated with these materials is given by Fthenakis et al.[1]. The best source for TLV (toxic limit values) and other specific hazard information is the MSDS (material safety data sheet) supplied by the manufacturer of the material.

The storage, use, and disposal of these materials will be reviewed. The primary concern of our Safety Committee and the management at Solarex involves the anticipation of errors (both mechanical and human) that could result in the uncontrolled release of any of these materials. The thorough investigation of accidents and near accidents provides insight into the weaknesses in the approaches taken. While some of the concepts described below are not new, we would like to share some of our experiences in the implementation of them.

MATERIALS AND PACKAGING

The feedstock materials in use at Solarex include silane, phosphine (concentration less than 5% in silane), diborane (< 5% in silane), and germane. Our policy for the packaging of these materials includes the use of only steel cylinders (for quantities >200 grams), which are fitted with a flow restricting orifice (0.010" diameter) in the CGA (Compressed Gas Association) connection.

We have decided that the claimed advantages of gas purity made for aluminum gas cylinders are offset by an increased risk. The risk includes the possibility that a leaking jet of silane in the area of the bottle could either melt or weaken a portion of the cylinder[2]. Also, there have been several products recalled by suppliers[3-5] of feedstocks in aluminum cylinders due to a fatigue cracking problem associated with specific lots of aluminum cylinders made by one manufacturer. A study performed by Linde Union Carbide[5] indicates that the fatigue cracking problem may apply universally to cylinders made from aluminum alloy 6351. Solarex does use some low content (<200 grams) aluminum cylinders in the research department.

Orifices (0.015" diameter), have been shown to reduce the flow of silane (500 psig) to the extent that well ventilated (500cfm through an 8" duct) gas cabinets did not suffer any damage due to explosion or severe temperature rise[6]. In a second study[7] by the same investigators, a 0.006" orifice was used to release silane into a ventilated cabinet without incident. It is clear that the flow-restricting orifice can serve to lessen the probability of a gas cabinet explosion. The orifice diameter that we have adopted has a cross sectional area of about half that used in initial Hazards Research study and three times that of the latter study. It is the smallest orifice size currently available to us. Some special considerations for the purging of CGA connections with flow-restricting orifices is given by Quinn and Rainer[8]. Gas cylinders of silane with and without flow restricting orifices fitted in the CGA connection were compared by a gas chromatography study of the purity of the gases delivered after a standard purging procedure had been carried out (purge techniques are described by Hardy and Shay[9]). The gases were anlayzed on a Hewlitt-Packard Model 5890. We found no increase in oxygen-containing species in the cylinders fitted with the orifices. We have not as yet experienced an incident of a clogged orifice.

It is our policy to adopt remotely operated pneumatic valves on these gas cylinders as soon as they become available since these can be easily interfaced with existing alarms and interlocks. Several pneumatic valve designs have been reviewed. Thus far we have identified one valve that incorporates the necessary backup systems to meet our concerns. Despite rumors that gas cylinders with pneumatic valves are available, we had not been able to obtain feedstocks that meet our purity specification in cylinders with these valves. We expect to receive a silane cylinder from MG Industries fitted with a pneumatic valve for evaluation soon.

Solarex does not yet have an outside storage facility for its dangerous materials, although a facility has been designed and should be ready in early 1988. To reduce the risk associated with gas storage, we have worked out a "just in time" delivery schedule with our suppliers. On delivery, all gas bottles are leak-checked for external leaks with hand-held detectors which are sensitive to a few ppm of silane in air. Any indication of an internal valve leak in a silane tank (pop when the dust cap is removed) results in the cylinder being returned to the vendor.

GAS CABINETS

Gases in use are kept in well ventilated steel cabinets (face velocities over 150 ft/min at the service door). The gas cabinets are fitted with low temperature (135°F) water sprinklers. A Telos (Telos Labs, Fremont, CA) gas detector constantly monitors all gas cabinets and areas where gases are used. A Telos gas detector probe port is fitted to the ventilation manifold that serves each group of cabinets. Some considerations for the choice of detection apparatus at Solarex are reviewed by Dickson[10]. A second level of detection is provided by smoke detectors which are sensitive to fire related to leaking silane or silane mixtures. Gas cabinet ventilation is both monitored and alarmed.

We are examining several options for constant germane monitoring. Germane poses a particular problem for photoionization, or flame photometric based detection (Telos employs this method) due to its high photon energies. Recently we ordered a solid state germane-silane detector (Model SP-1802-P) from Matheson Gas Company which should have the necessary sensitivity to detect small germane leaks[11]. Until on-line germane detectors are installed, we are purchasing germane in small quantities (<200 grams per cylinder).

REACTOR DESIGN CONSIDERATIONS

Like design considerations elsewhere, reactor design is aimed at the prevention of uncontrolled accumulation and release of dangerous materials. It is our belief that reactors can be designed to compensate to a large extent for both human error and component malfunction. The Solarex machine certification (some aspects of the criteria are currently under review) criteria is shown in Table 1.

Item #1 prevents the unintentional release of a toxic gas through operator error or a pumping failure. Note, this interlock should not use the same pressure sensor that the process throttle valve uses. Because, a sensor failure could result in both a pumping failure (due to a closed throttle valve) and an interlock failure. The second item provides a safe guard against electrocution during cleaning and maintenance of the apparatus. Solarex employs cryopumps on many deposition systems for the vacuum preparation of substrates prior to deposition. The cryopump is not used for the pumping of toxic materials. It is conceivable that a combination of valve failures could lead to pumping (storage) of the process gases. This problem is characterized by a temperature rise in the cryopump (item #3).

Table 1

Machine Certification Criteria

1. Process gas interlock to prevent chamber pressure >2.0 torr.
2. High voltage interlock to prevent high voltage discharge when chamber is open or at high gas pressure (>2.0 torr).
3. Temperature increase interlock on cyropumps to prevent the freezing and accumulation of process gases.
4. Purge assemblies which have inert gas inlets with one-way valves.
5. Purge manifold assembly pressure (and vacuum) gauges.
6. No high pressure (cylinder side of pressure regulator) routing to vacuum equipment (such as cross tee purge).
7. Emergency shutdown procedures posted on the control panel.
8. Exhaust dilution gas interlock.
9. Scrubber over temperature and pressure interlocks.

Purge assemblies (if used) deserve special consideration since this operation is performed manually and it normally by-passes the deposition system with its interlocks. The only plumbing on the high pressure side of the gas regulator is an inert gas inlet fitted with a one-way valve to prevent the back diffusion of process gases into the purge lines in the event that the process gas cylinder valve has failed (we have experienced cases where a cylinder valve would not close) or is not correctly closed during the purging. A vacuum manifold fitted with vacuum and pressure gauges clearly visible at the location of the valves needed for purging is used. A purge assembly is shown in Figure 1.

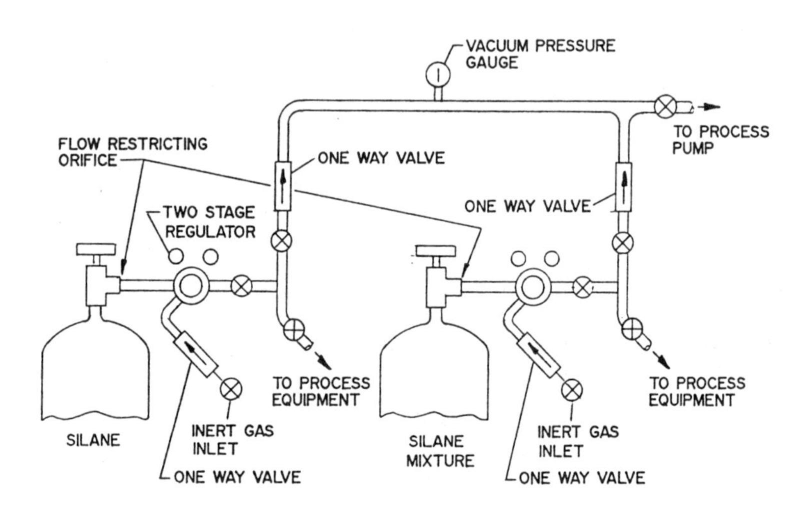

Figure 1. Process gas assembly with inert gas inlet and vacuum manifold.

WET SCRUBBER SAFETY CONSIDERATIONS

Solarex has chosen to use wet scrubbers to reduce the quantity of toxic materials released into the atmosphere. Scrubbers were chosen over thermal decomposition because they require less frequent cleaning and neither store nor release toxic solid materials. Scrubbers can be very effective (>99% efficient) in the removal of HF and $SnCl_4$ from the effluent of our tin oxide glass coating process. Both feedstocks and the effluent of the tin oxide glass coating process pose a much lower threat than those used for the amorphous silicon deposition.

Scrubbers are somewhat less efficient in the removal of silane from the exhaust of the silicon deposition equipment. On the negative side, scrubbing efficiency is a function of dilution (in our case dry nitrogen is used as the diluent). We are currently using a 2% NaOH solution (pH 13.7) scrubbing solution. Increased silane concentration in the exhaust represents an increased hazard should an accidental release occur, while an increased dilution (lower silane concentration) reduces the efficiency of the scrubber. The scrubber (Airprotek model SD201) silane removal efficiency can be as high as 95% when silane flows of ~1.0 SLM (standard liter per min) are diluted with ~3.0 SLM of nitrogen[12]. The number of deposition systems manifolded to a scrubber and the maximum silane flow rates they employ has been adjusted such that each scrubber has an inlet silane flow rate <1.0 SLM.

We conducted a limited study of scrubber efficiency using an alternate solution consisting of sodium phosphate tribasic at pH12. Our measurements indicate that the efficiency using this solution is much lower than those obtained by Airprotek[12] using NaOH solutions. The results of this study were not sufficiently promising to warrant a more thorough investigation at this time. Currently, only NaOH solutions are being used in our scrubbers.

Our analysis of the effluent of glow discharge deposition indicate that a large portion (>70%) of the silane is not consumed by the amorphous silicon deposition process. The deposition process generates some hydrogen approximately equal to twice the amount of silane consumed through the glow discharge decomposition of silane. It is unlikely that the concentration of silane in the exhaust system (between the deposition system and the scrubber) and in the scrubber are low enough to be below the flammability limits (~0.5% in inert gas) for silane-hydrogen mixtures[13-15]. It is necessary to consider any air leak into this exhaust plumbing to be very serious. Any entrance of air into the exhaust plumbing as well as inefficient scrubbing is characterized as a temperature rise in either the scrubber inlet, the scrubber itself, or the scrubber exhaust. We have prepared a temperature alarm package for our scrubbers to augment the pressure alarm package available through the manufacturer. We presently are using 130°F as the alarm point. The positions of the thermocouple probes are shown in Figure 2. The scrubber exhaust is not inert as it contains additional hydrogen since silane is decomposed to hydrogen by the scrubber in accordance with reaction given below where the final products are a complex silicate polymer and hydrogen.

$$SiH_4 + 4NaOH \rightarrow (Na_4SiO_4)_x + 4H_2 \qquad (1)$$

The scrubber exhaust is diluted further with dry nitrogen to reduce the flammability hazard.

A flame where the scrubber exhaust enters the atmosphere (outside the facility, on the roof) is an indication of a problem. There appears to be two processes by which a pyrophoric gas is exhausted from the scrubber. First, when a large inert gas dilution is used on the process exhaust (scrubber inlet), poor scrubbing efficiency results in a sufficient silane concentration in the scrubber exhaust to ignite the silane/hydrogen mixture. Secondly, inadequate dilution of the scrubber exhaust could lead to a pyrophoric hydrogen/silane mixture being exhausted.

TIN OXIDE COATINGS

Solarex employs a Watkins-Johnson CVD furnace for the tin oxide coating of glass to provide substrates for the solar cell operations. Initially, tetramethyltin was used as the tin-containing feedstock. The use of tetramethyltin as a feedstock for our tin oxide coatings has been suspended since early 1986 as a result of two factors. One, the CVD furnace was found to leak small quantities of tetramethyltin from the areas around the injector heads (it is our understanding that the injector heads have been redesigned to prevent leakage[16]). It is our policy to minimize any known amount of a toxic substance in our work place even if the values do not exceed TLV's. Secondly, we found that superior coatings could be prepared using less dangerous feedstocks (tin tetrachloride).

EMERGENCY RESPONSE

Alarm panels for the Telos gas detectors are located at the location of the probe and at a general alarm panel located in the office area of the facility. Smoke detectors trigger the facility fire alarm with locations being indicated on the general alarm panel. Sprinklers also trigger the building fire alarm. In the event of an alarm, the emergency response team consisting of technical members of the Safety Committee and management meet at the general alarm panel. SCBA's (self contained breathing apparatus), hand-held leak detectors, a library of MSDS', and protective clothing are located at the alarm panel. All employees exit the building from the exit nearest their work station and then gather outside the facility in one central location. This facilitates taking a head count. Anyone with knowledge of the problem re-enters the facility near the alarm panels to advise the Emergency Response Team.

TRAINING AND EDUCATION

All personnel receive training about the hazards associated with the production of amorphous silicon and the emergency response procedures. Only a few individuals are trained and certified to perform work on gas-handling systems (includes cylinder changing and leak checking). Cylinder dust caps are removed only after the cylinder is placed in the ventilated gas cabinet. Two technicians are required to perform this operation and must be equipped with both SCBA and hearing protection (the pop heard when silane is released during dust cap removal from a faulty cylinder can cause temporary or permanent hearing loss). The associated connections are carefully leak-checked prior to the cylinder valve being opened. No work is carried out on hot (charged with silane) gas-handling plumbing. We are presently considering the automation of manifold leak checking, purging and cylinder-changing procedures.

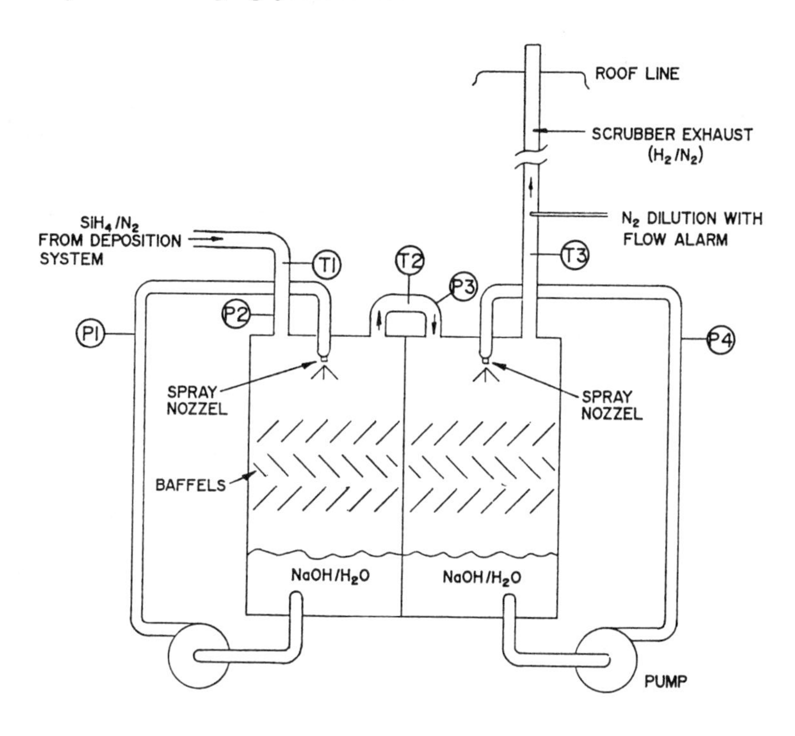

Figure 2. Cross-section of two-stage scrubber showing the location of thermocouple probes (T1, T2 and T3) used for our in-house designed-over temperature alarms. P2 and P3 are pressure sensors to detect gas line blockage. P1 and P4 ensure proper solution pressure at the spray nozzle.

SUMMARY AND CONCLUSIONS

The safety measures that we feel are necessary and practical at this time have been reviewed. We have attempted to identify problem areas associated with equipment using dangerous materials. The flow restrictors, manifolding considerations, interlocks and sensors that prevent the progression of an error to a dangerous situation were reviewed. In particular we feel that scrubbers should be equipped with temperature alarms so that early detection of problems can take place. At this time there appears to be no reason to accept the possible risk associated with aluminum cylinders for silane and related materials.

We would like to encourage the gas vendors to proceed with their plans to employ pneumatic valves on their gas products. It is our intention to require pneumatic valves in the near future. The pneumatic valves are to be directly interlocked with the building fire alarms, gas detectors, and machine controls.

ACKNOWLEDGEMENTS

The authors would like to thank M. Finn (Industrial Hygiene Department) and R. Golec (Safety Department) of the Amoco Environmental Affairs and Safety Department for many important suggestions regarding the handling of dangerous materials and for funding the design engineering of our gas cabinet ventilation system.

REFERENCES

1. V.M. Fthenakis, P.D. Moskowitz, and L.D. Hamilton, Solar Cells, Vol. 19, Nos. 3-4, p.369 (1987).

2. S. Tunkel, private communiation, Hazards Research Corp., 200 E. Main St., Rockaway, NJ 07866-3694.

3. Airco Products, Murray Hill, NJ, product recall dated 8/11/86.

4. Alphagaz Company, Walnut Creek, CA, product recall dated 7/31/87.

5. Union Carbide Corporation, Linde Div., Morrestown, NJ, product recall and safety notice dated 5/18/87.

6. Rep. 5038 from Hazards Research Corporation, Rockaway, NJ to IBM, Essex Junction, VT, 5/11/82.

7. Rep. 5488 from Hazards Research Corporation, Rockaway, NJ to IBM, Hopewell Junction, NY, 12/9/83.

8. W.E. Quinn, D. Rainer, Solid State Technology, vol. 29, no. 7, pp.63-66, (1986).

9. T.K. Hardy and R.H. Shay, Solid State Technology, Vol. 30, no. 10, p.13 (1987).

10. C.R. Dickson, Solar Cells, Vol. 19 nos. 3-4 (1987).

11. Matheson Gas Co., P.O. Box 85, E. Rutherford, NJ, technical release (1987).

12. Airprotek, 1101 S. Winchester Blvd, Suite C-130, San Jose, CA, technical release (1987).

13. P. Bachmann and H.P. Berges, Solid State Technology, vol. 29, no. 7, p.83 (1986).

14. Messer Griescheim, "Product List for the Electronic Industry" (1980).

15. M.L. Hammond, Solid State Technology, vol. 23 (12), p.104 (1980).

16. B. Mayer, Watkins-Johnson Co., Palo Alto, CA, correspondence dated 6/4/86.

SAFETY CONSIDERATIONS IN USING TOXIC AND EXPLOSIVE GASES IN A PRODUCTION ENVIRONMENT

K. Hoffman, S. Brubaker, P. Nath, G. DiDio, M.Izu, and J. Doehler
Sovonics Solar Systems, 1100 W. Maple Road, Troy, MI 48084
(subsidiary of Energy Conversion Devices, Inc.)

ABSTRACT

This paper will detail considerations employed to insure the safe use of hazardous gases in a thin film manufacturing facility operated by Sovonics Solar Systems. The gas handling equipment will be described as well as practices and procedures that have been developed.

INTRODUCTION

Thin film, amorphous silicon alloy based photovoltaic power modules are manufactured at Sovonics Solar Systems with a roll-to-roll photovoltaic processor[1]. Among the gases used by this machine are silane, hydrogen, argon, silicon tetrafluoride, and phosphine, which display varying degrees and types of hazards, such as toxicity, flammability, liability to suffocate, and, under certain conditions, liability to spontaneously explode. Further, the roll-to-roll deposition reactor is operated 24 hours per day, 7 days per week. These facts combined constrain the design of a safe gas handling system.

The gas handling system may be thought to consist of four functional areas: cylinder storage, source delivery/distribution, deposition processor, and exhaust pumping with effluent emission. The cylinder storage facility houses cylinders, either full or empty. The source delivery area consists of the gas handling components and associated equipment to deliver gas from in-use cylinders to the deposition processor. The deposition processor receives process gas at low pressure for use in the deposition process cycles. The remainder of the gas handling system includes vacuum pumps and exhaust conditioning equipment. Although most of the discussion below will consider each of these areas separately, it will be apparent that this gas handling system was designed as an integrated unit.

GENERAL DESIGN CONSIDERATIONS

In order to insure the quality of construction and to develop the in-house expertise, the gas system was designed and constructed by Sovonics. The design was based on recommendations from gas and equipment manufacturers[2] and consultants[3], and on design ideas in various articles[4-6]. The design was finalized only after a careful hazard assessment[7-11].

DESIGN CONSIDERATIONS - CYLINDER STORAGE

Recommended practices for the storage of cylinders have been documented[6,12-14]. Consequently, a gas storage facility was constructed as an outdoor, isolated area. This area is fenced with barbed wire, has a sloped concrete floor and a roof. The facility is marked with safety signs identifying specific hazards. All process gas cylinders (except hydrogen) are segregated by hazard type and stored there after receipt from vendor. Hydrogen is stored separately in a similarly constructed facility, which includes purging and pressure regulating equipment (see below).

DESIGN CONSIDERATIONS - SOURCE DELIVERY AND DISTRIBUTION

The source delivery with its dedicated purging system was designed and built. A separate, highly automated, microprocessor controlled, and remote source delivery system was designed for hydrogen (figure 1). A fundamental decision that had to be made for the source delivery design was the degree of automation appropriate for this gas handling system. The advantages of an automated system include: less opportunity for operator error, easier operation (operator training need not be as extensive), and practicality of making the system more physically remote. In contrast, advantages of a more manual system include: ability to use the complex human brain, greatly reduced number of potentially unreliable components and fittings, reduced cost, and increased flexibility to process changes. Other considerations include the component technology available at the time, competence and experience of gas system operators, required processor capabilities, and availability of programmers. The hydrogen delivery system chosen was based on a commercially available configuration[15]. This unit will provide experience using a heavily automated delivery source and remove a significant potential explosion source from the building.

All connections on the gas system were welded wherever possible except on discrete components. All non-welded connections, bar the CGA fitting-cylinder interfaces, were attached using VCR[16] fittings with metal gaskets. Welds were made using a microprocessor controlled orbital TIG welding machine. Test welds were fabricated and cut open frequently during construction to insure proper welder operation. Regulators were chosen whose internal construction provide a positive shut off should normal regulator action fail. Pneumatic valves were installed upstream of every process gas regulator and some cylinders are equipped with pneumatic cylinder valves. The pneumatic valves are operated by transducers which are electrically in series with a normally closed solenoid. An interlock closes the pneumatic valves on the source delivery when one of many conditions are not met. The manual valves have visual position indicators.

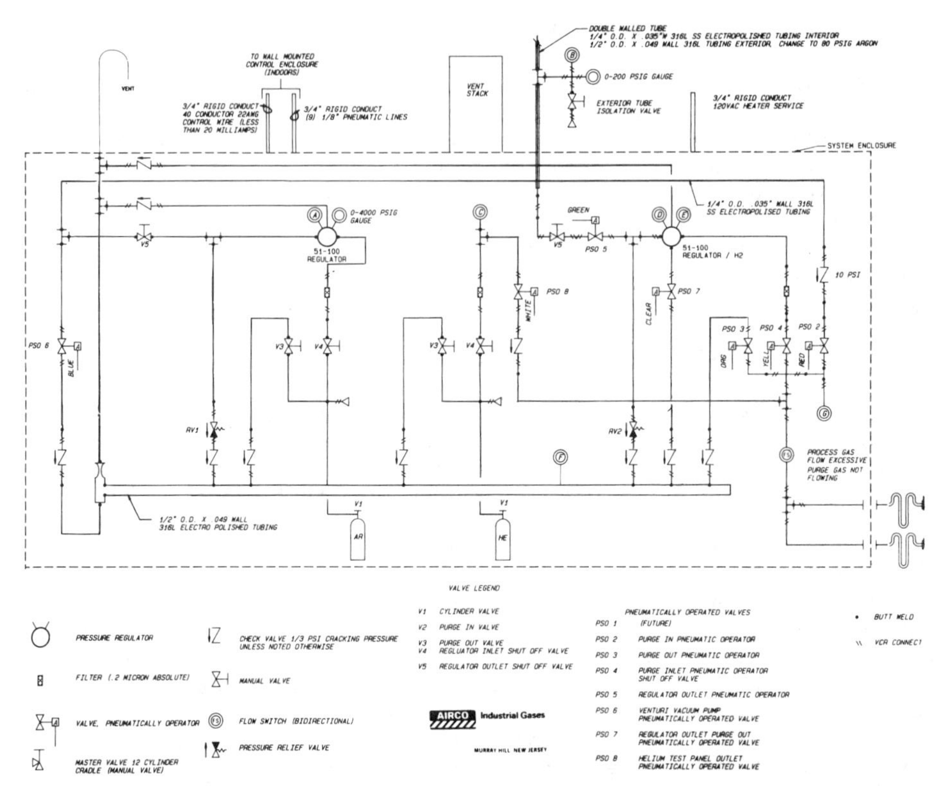

Figure 1. Hydrogen source delivery and purge system.

It can be noted in examining figure 1 that the purge source has the capability of delivering high pressure inert gas to all areas in the source system which have high pressure process gas in them during operation. An important part of this design is to be able to check for leaks at pressures higher than normal processing pressures to insure integrity of the tubing and gas handling components.

The slowness of the diffusion of gases at atmosphere is the reason why it is practically necessary to use a vacuum generator. For example, at room temperature and 30 psig it will take more than 15 minutes for most air atoms to diffuse 10 centimeters into argon! Thus, dilution purging is much more effective when the gases are allowed to mix thoroughly; the "double pigtail" design of the purge inlets for each process gas further enhances the purging efficiency[17]. Although normal operation dictates that gases will be removed from process lines at the end of the deposition cycle through the deposition processor, a capillary bypass was installed leading to the vacuum generator so that hazardous gases can be removed from lines slowly. Even in an emergency this will insure that gases will be highly diluted before releasing them to the atmosphere.

There is a network of flowmeters and pneumatically operated valves which distribute gas to the deposition processor. These components and associated fittings are located in a purged gas cabinet. This cabinet is always closed during deposition processor operation. Double-walled tubing is used to transport hydrogen gas to the flowmeters. As can be observed in figure 1, both inner and outer tubes are continuously monitored for leak integrity.

DESIGN AND CONFORMANCE TESTS - DEPOSITION PROCESSOR

The deposition processor has multiple interlocks to protect operators. For example, hazardous gases are prevented from flowing to the machine unless many conditions are met. There are also interlocks for hazards such as electrical shock. A detailed description of the interlocks on the deposition processor is beyond the scope of this paper. However, it was found to be important to model the interlocks and machine control sequences on a computer using ladder logic. This enabled an accurate review of the interlocking design to take place as well as providing a training tool for operators. In addition, a dedicated computer is interfaced with the machine to monitor controls and interlocks providing a data base of processor operation. Vacuum apparatus hazards with guidelines for management of them have been reviewed[4,5,18].

DESIGN CONFORMANCE TESTS - SOURCE DELIVERY

Through the use of the high pressure purging source (figure 1), every gas line, weld and component upstream of the regulators was tested for leaks using helium at more than 2000 psig. Helium was detected using a standard helium mass spectrometer equipped with a sniffing probe. Also, valves were checked for through leaks by evacuating and sampling one side of each valve while exposing high pressure helium to the other. Gas lines and components between the regulators and the deposition processor were tested under vacuum using helium as a tracer gas enveloping items under test.

The above leak checking was performed twice for each area to insure that mistakes were not made. A labeling system was developed to control this. Also, leak checking equipment calibrations were frequently verified. Each weld and fitting was reexamined visually for imperfections.

DESIGN CONSIDERATIONS AND PROVE-OUT PUMPS AND EXHAUST EFFLUENT CONDITIONING

The principles and equipment used to safely pump large quantities of hazardous gases have been established[3,19,20,21]. It is important that extensive gas flushing be used, including inert gas purging at filters and pump ballasts, over oil reservoirs, and into the exhaust line. Controls to insure that purging is adequate were interlocked to the deposition processor operation. That is, the processor is unable to use hazardous gases unless proper purging is in effect. Exhaust handling equipment is located in well ventilated areas which are not part of the normal traffic of the building.

The use of oxygen sensors to monitor the exhaust areas is highly recommended to insure that exhaust lines have been properly purged prior to the introduction of hazardous gases and to monitor the leak integrity of the exhaust lines. This is especially important in processes where both silane and hydrogen are present as the concentration for pyrophoric ignition is much lower[4]. It is possible to use oxygen sensors cheaply and reliably[22]. The sensors chosen have a lower detection limit of 0.1 at. % and a 6-12 month lifetime. Because their cost is low, these sensors have been liberally installed throughout the exhaust system and have been instrumental in locating suddenly appearing leaks (such as cracked bellows).

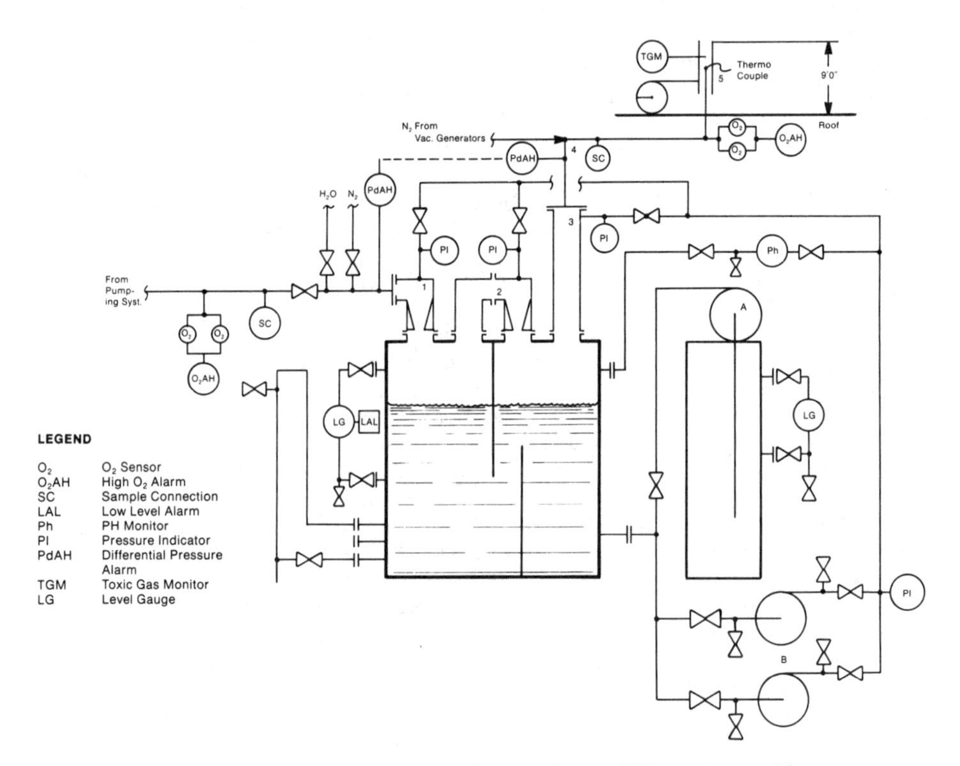

Figure 2. Controls and monitors for effluent scrubber.

Emission conditioning is performed by exposing the waste gases to an effluent scrubber and diluting them before ejecting them to atmosphere. The safe use of a scrubber is enormously enhanced by controls and monitors (figure 2). Referring to the figure, exhaust gas enters the scrubber by Venturi pump action at point (1), flows by caustic sprays at (2), flows through a "packed column" at point (3), and is finally ejected to atmosphere at (5). The pH value of the caustic solution is maintained for optimal reaction. Sensitive differential pressure sensors are used to detect blockages before exhaust pressures rise to unsafe levels. Oxygen sensors continuously monitor leak integrity. Pressure gauges and level indicators are used to monitor the caustic fluid. A dedicated hydride monitor (point 5) constantly monitors scrubber performance. Measurement of emissions from this conditioning system show that all gases are less than their TLV's[10] as they leave the stack. Further, analysis indicates[8] that the maximum expected ground level concentrations should be many orders of magnitude less than the TLV for each gas.

OPERATION PROCEDURES

In order to maintain a safe program for using hazardous gases periodic review of design and operating procedures is essential. This is so because state-of-the-art in gas handling technology is constantly improving, operational experiences accumulate uncovering design or operation flaws, and process changes may disturb designed strategies. It is important to get persons interfacing directly with gas handling equipment to provide feedback[23].

Emergency procedures, which describe potential emergencies and how they are detected, detail planned responses, and which indicate communication networks, have been written. Responses (for example, evacuation exercises) are practiced so that all employees are instructed and can respond with intelligent confidence to emergencies. First aid kits[24], eye wash stations, showers, fire fighting equipment, self-contained breathing apparatus, stretchers, portable hazardous gas monitors, and protective clothing are available throughout the factory. Written escape routes and alarm legends are posted in various locations. Material safety data sheets are maintained along with lists of hazardous materials inventories. These are all part of the emergency preparedness plan[25] for this facility.

Cylinder inspection is performed for incoming cylinders[2]. Written cylinder handling rules are obeyed by personnel involved in handling the cylinders. Only trained personnel are allowed to handle bottles. On site inventories are kept to a minimum by arrangement with gas vendors. Technical training for changing cylinders and operating the purge system also includes teaching of operation principles. Aids for training include cylinder change procedures and written outlines of the gas system design principles.

Equipment which handles hazardous gases at pressures greater than atmospheric is periodically leak checked using methods detailed above. The pumps and scrubbing equipment are also routinely maintained. Gas monitoring components are recalibrated as required to insure correct performance. All maintenance is documented.

REFERENCES

1. P. Nath et al. in Technical Digest of the International PVSEC-3, Tokyo, Japan, 1987 p.395
2. "Guide to Safe Handling of Compressed Gases", (Matheson, Division Searle Medical Products, U.S.A., Inc., 1982)
3. private communication (Systonics, Inc., Winchester, MA 1983)
4. V. Fthenakis, P. Moskowitz, and J. Lee, Solar Cells 13, 43 (1984)
5. M. Hammond, Solid State Tech., (Dec. 1980), p.104
6. L. Fleur, Microelectronic Manufacturing and Testing, (Nov. 1983), p.53
7. V. Fthenakis, P. Moskowitz, and L. Hamilton, Solar Cells 19, 269 (1986-7)
8. P. Moskowitz, V. Fthenakis, L. Hamilton, and J. Lee, Solar Cells 19, 287 (1986-7)
9. M. Sittig, "Handbook of Toxic and Hazardous Chemicals and Carcinogens" (Noyes Publications, Park Ridge, N.J. 1985)
10. "Threshold Limit Values for Chemical Substances and Physical Agents in the Workroom Environment" (American Conference of Governmental Industrial Hygenists, Inc., Cincinnati, OH 1985)
11. F. Mackison, R. Stricoff, and L. Partridge, "NIOSH/OSHA Pocket Guide to Chemical Hazards", Rep. DHEW(NIOSH) Pub.No. 78-210 (Washington, D.C. 1978)
12. "Guide to Safe Handling of Compressed Gases" (Matheson, Division Searle Medical Products, U.S.A., Inc., 1982) pp.63-77
13. "Safe Handling of Compressed Gases in Containers" CGA Pamphlet P-1, (Compressed Gas Association, Inc., Arlington, VA 1984)
14. R. Collins and E. Flaherty, Solar Cells 19, 355 (1986-7)
15. private communication (Airco Special Gases and Equipment, Delran N.J. 1987)
16. registered trademark (Cajon Company, Macedonia OH)
17. T. Hardy and R. Shay, Solid State Tech., (Oct. 1987) p.131
18. L. Beavis, V. Harwood, and M. Thomas "Vacuum Hazards Manual" (American Vacuum Society, N.Y. 1979)
19. "Pumping Hazardous Gases", D. Fraser, ed. (American Vacuum Society, N.Y. 1980)
20. P. Bachman in Safety Aspects of Effluents from CVD Process Systems, Semiconductor Equipment and Materials Institute, Inc. Technical Education Program (San Francisco, CA 1986)
21. J. O'Hanlon, "A User's Guide to Vacuum Technology" (John Wiley and Sons, N.Y. 1980) p.323
22. G-CELL 33-200 Specification Sheet (G.C. Industries, Chatsworth, CA 1985)
23. J. Veltrie, Solar Cells, 19 (1986-7) 337
24. W. Breaker, A. Mossman, and D. Siegel, "Effects of Exposure to Toxic Gases - First Aid and Medical Treatment" (Matheson, Lyndhurst, N.J. 1979)
25. R. McConnel, Solar Cells, 19 (1986-7) 345

INTERACTION OF SAFETY AND THE FACILITY FOR PHOTOVOLTAIC R & D

R. P. Gale, J. P. Salerno, P. M. Zavracky, and W. P. Brissette
Kopin Corporation, Taunton, Ma. 02780

ABSTRACT

At Kopin there is a program to develop high efficiency solar cells for terrestrial and space applications. The work centers around GaAs/AlGaAs structures grown by organometallic chemical vapor deposition (OMCVD) and involves a range of thin-film and metallization operations in order to fabricate complete solar cells. These processes use hazardous materials. In order to safely run these processes, we have integrated safety features into the facility through the overall physical plant design and placement of equipment. Handling of chemicals and waste, and the safety features of the facility at Kopin are described.

INTRODUCTION

To more easily visualize the different safety issues, the solar cell fabrication process has been divided into three general steps: wafer cleaning, cell structure deposition, and cell processing. The waste streams for each of these steps are shown schematically in Fig. 1. The feedstock for each of these steps is solvents and acids for wafer cleaning, flammable liquids, flammable gases and toxic gases for cell deposition, and solvents, acids, and bases for cell processing. In addition to completed solar cells, the process produces solvent waste, acidic waste, toxic waste, and toxic gas. All of these are handled appropriately by a solvent recovery system, an acid neutralization system, toxic waste isolation and collection, and a waste gas scrubbing system.

The remainder of the paper is organized into three sections. The first section is on feedstock handling, including the receiving, storage, and transport of the raw materials used in the process. The second section discusses our waste handling systems listed above. The third section describes our facility safety systems, which include spill control, process alarms, and fire alarms.

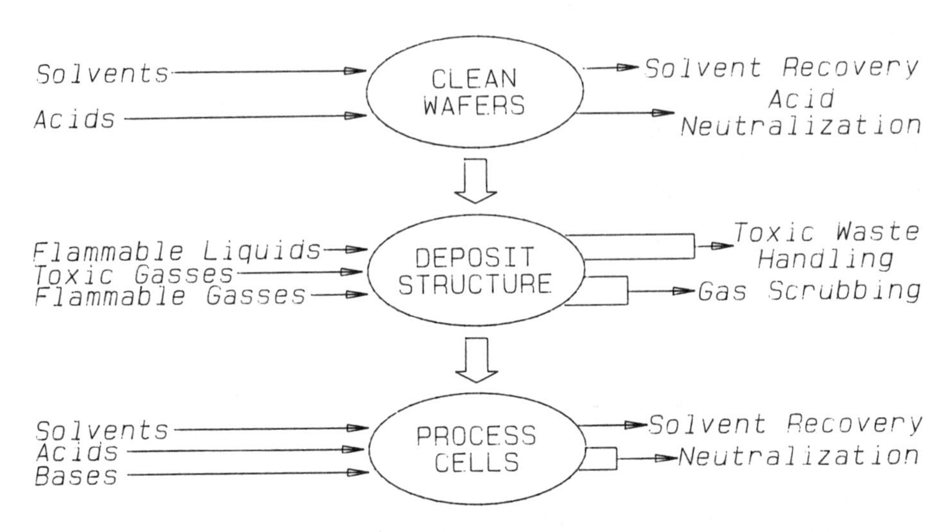

Fig. 1. Waste streams for each of the solar cell fabrication steps.

FEEDSTOCK HANDLING

Feedstock handling for solar cell fabrication is summarized in Table I. Upon receipt, and while the trucker is waiting, all solvent, acid, and base packages are inspected for container integrity. If there are any signs of leaks or spills, the entire lot is segregated and the vendor is notified. Upon acceptance, the chemicals are stored in vented, fireproof cabinets with separate cabinets for solvents, bases, oxidizers, and acids (except HF which is stored separately). The chemicals are transported in a cart with polypropylene wells for each one-gallon bottle. The bottles are well protected and spills from broken bottles would be self-contained. The chemicals are used in hoods which are designated as solvent or acid/base. Two exhaust systems were installed for chemical hoods, one for the solvent hoods and one for the acid/base hoods. Any potentially hazardous mixing of the two chemical groups is thereby avoided. In addition, separate drains are used in each type of hood, eliminating the mixing of waste streams. Details of these sinks are given in the next section.

The only flammable gas used in appreciable quantities is hydrogen. Hydrogen is stored as gas outside the building in a permanent, fenced-in sled of high-pressure cylinders. Delivery of hydrogen is made from a truck directly to this storage facility. Hydrogen gas is run

Table I. Summary of feedstock handling

Feedstock	Receiving	Storage	Transport	Use
Solvents	Inspect container	In vented fireproof cabinet	Chemical-resistant cart	In solvent hoods
Acids and Bases	Inspect container	In vented fireproof cabinets	Chemical-resistant cart	In acid hoods
Flammable Gas	Outdoor delivery	Outdoor Storage	Welded, stainless steel lines	In vented systems in monitored locations
Flammable Liquids	Inspect in vented hood	Fireproof packed in fireproof cabinet	Fireproof packed and sealed	In vented systems in monitored locations
Toxic Gas	Leak test cylinder	In vented cabinet	On gas cart via access corridor	In vented cabinets in monitored locations

into and through the building in welded, stainless steel lines. The length of lines has been minimized by locating all hydrogen-using systems in close proximity to the hydrogen storage facility. A positive shut-off valve located in the hydrogen line outside the building allows automatic shutoff of the hydrogen supply to the building and accompanying nitrogen purging of all the hydrogen lines within the building. This shutoff valve is controlled by the process alarm system. Hydrogen is used in ventilated systems in work areas monitored by flammable gas detectors, which are also part of the process alarm system.

Flammable liquids handled include trimethylgallium and trimethylaluminum, both of which are used in the organometallic chemical vapor deposition (OMCVD) process. These liquids are contained in small unpressurized, stainless steel cylinders. The cylinders are shipped and packed in metal cans with fireproof, absorbant material. The cans and cylinders are inspected upon receipt, and stored in a fireproof cabinet. Cylinders are transported inside the cans, and used in ventilated systems in locations monitored by smoke and heat detectors.

Toxic gases are handled with specific procedures by personnel knowledgeable in their properties and hazards. The gas cylinders are tested for leaks with a portable leak detector while the trucker is waiting. If there is any sign of a leak, the vendor is notified. Cylinders are stored in vented, sprinklered cabinets, and transported on gas cart via access corridors. The cylinders are installed using vented cabinets in locations monitored for gas leaks. The monitoring system is part of the process alarm system.

WASTE HANDLING

Waste handling begins at point of use of the feedstock materials. It therefore maintains the same isolation between chemicals described previously, and uses multiple but simplified treatment systems. These systems include solvent collection, water neutralization, and toxic material collection. Each of these systems addresses a single operation and type of waste material. Each of these systems has been integrated into the facility for highly reliable operation, and ease of use and maintenance.

The solvent collection system is based on the use of individual sinks in every solvent hood for the disposal of three main organic solvents used, alcohols, acetone, and trichloroethylene. The latter sink is also used to collect organics not in the former two categories. The drain from each sink is run into the corresponding drain which runs the length of the building in a concrete trench under the clean room floor, shown schematically in Fig. 2. The trench connects to a pit where the drains empty into three corresponding collection drums. When solvent levels in the drum reach a certain point, the solvent is pumped up into a disposal drum. The material in the disposal drum is treated as hazardous waste and picked up by a registered waste handler.

The trench and pit are key elements to the solvent collection system and the water neutralization system. The trench is monitored for moisture to detect leaks, and acts as a secondary containment vessel. The systems are gravity fed, making them simple to install and modify yet very reliable. The lines are accessible and avoid the problems associated with buried pipes. Lastly, the solvent collection system as part of the facility is simple to use, and works well.

The water neutralization system also makes use of the trench and pit. The system is shown schematically in Fig. 3. The sinks in the acid/base hoods all drain to a

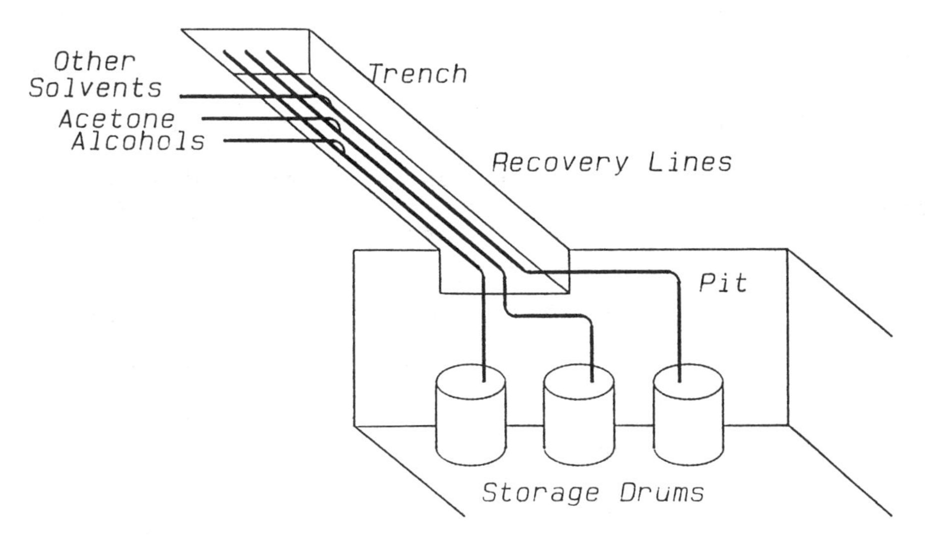

Fig. 2. Schematic diagram of the solvent recovery system.

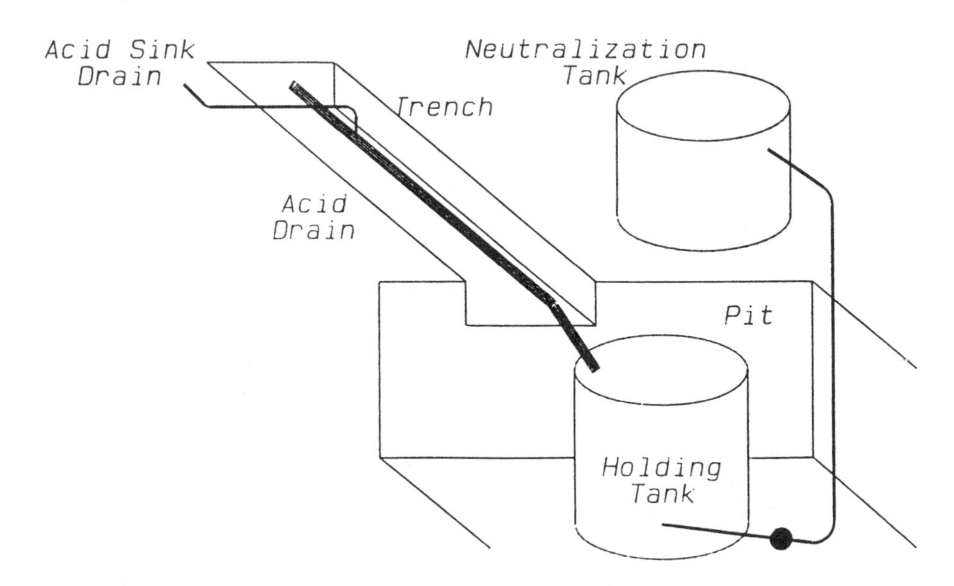

Fig. 3. Schematic diagram of the water neutralization system.

common line in the trench, which in turn empties into a large holding tank in the pit. When the water level in this tank reaches a predetermined level, it triggers an automatic sequence of events. The contents of the tank are first pumped up into a neutralization tank where the water is stirred while acid or base is added. The pH is continuously monitored and used to control the amounts of neutralization chemicals added. After reaching and maintaining a pH of about 7 for a period of time the tank contents are fed into the municipal water drain. At regular intervals, samples of the water are taken and tested at an independent laboratory.

Toxic material waste is handled in yet another, separate system. The deposition system cleaning operation removes coatings of arsenic and gallium arsenide in a cleaning vessel in a designated hood. Solutions containing arsenic are poured into a special drain which runs into a collection bottle. The bottles are packaged in drums, and treated as liquid hazardous waste by the contracted waste handler.

FACILITY SAFETY SYSTEMS

The facility safety systems monitor and/or address problems with the normal operation of the facility. These systems include spill control, process alarms, and fire alarms, and are independent of monitors which are located at the various waste treatment systems. Procedures and training are both an integral part of these systems.

Spill control equipment is located in each room or work area where chemicals are used or stored. Amounts of spill control materials are commensurate with the amounts of chemicals stored in the area, and containers for the disposal of used material are readily available. Material safety data sheets for those chemicals are kept in a binder in each room. In addition, a mobile spill cart is kept outside the clean room near the chemical storage area.

The process alarm system monitors the GaAs deposition equipment area and is designed to protect personnel rather than equipment. It includes smoke detection, hydrogen detection, and toxic gas detection, and controls hydrogen shutoff, exhaust fan speed, and both local work area and building evacuation. The detectors are primarily positioned in the rooms where personnel may be working, rather than in the equipment. The first and most sensitive detection level results in a local alarm evacuating the immediate work area and automatically increasing local exhaust fan speed. The higher level, or critical alarm, results in building evacuation and automatic shutdown and purging of the hydrogen system.

In addition to the process alarm system, there are a number of detectors on equipment and in systems and ducts to monitor for equipment failure. These detectors will result in a buzzer and light flashing, with some shutting down of individual pieces of equipment. They are not, however, personnel monitors and are therefore not integrated into the process alarm system.

The facility also has an independent fire alarm system. The fire alarms are activated by pull stations or operation of the building sprinkler system. The alarm causes building-wide evacuation horns to sound, calls the local Fire Department, and directs the process alarm system to shut off the hydrogen supply to the building. In the event of building evacuation by either fire alarm or process alarm, an emergency response team reacts to the situation.

SUMMARY

The systems for the handling of chemicals and waste have been integrated into the facility at Kopin. By keeping the various materials and waste streams separated, solutions to individual treatment and handling problems have been implemented in a simple and cost-effective manner. The operation of the facility to develop high efficiency solar cells for terrestrial and space applications is, in turn, enhanced.

A CRITICAL CONSIDERATION OF GAS SAFETY SYSTEMS AND COMPONENTS*

John J. Bordeaux
Safe Flow Systems, Inc., San Francisco, Calif. 94114

INTRODUCTION

There are three major subjects to be covered by this text:

A. Current safety code status for gas cabinets and gas delivery systems.
   1. California Assembly Bill - AB 1021 - status.
   2. Uniform Fire Code - Article 80 - revision status.

B. Current status of systems for hazardous gas applications.
   1. Requirements for a good system.
   2. Manual or automatic.
   3. Design variations.

Gas safety might very simply be divided into two distinct categories of concern: "Gas going into the system" and "Gas coming out of the system." The concentration here will be on "Gas going into the system." There is great certainty about the properties of those gases going into the system. They are stored in their own containers and the chemical nature can usually be well defined. It is not as easy to obtain the composition of all of the reacted species that exit these systems.

C. Current status of components used in Gas Safety systems.
   1. CGA Fittings and Cylinder Valves.
   2. Filters.
   3. Tubing.
   4. Fittings.
   5. Regulators.
   6. Valves.
   7. Specialty Components.

GENERAL DISCUSSION

A. Current safety code status for gas cabinets and delivery systems.

1. California Assembly Bill, AB 1021, Toxic Gas Model Ordinance.

A toxic gas model ordinance report has been developed for the California State Legislature by the Santa Clara County Fire Chief's Association. This document, called the AB 1021 Report, was prepared during 1986 and 1987 by a committee appointed by the Fire Chief's Association. If the document is to go further, it will be considered by the State Legislature in the form of a Bill. Whether or not it will become law in

* Any data, calculations, formulas, conclusions, or material abstracted from the material presented and/or taken out of context, in whole or in part, will be at the users risk and will not necessarily have the authority or the approval of the author or Safe Flow Systems, Inc.

California in its present form is pure conjecture, but it will face much debate when it gets to the floor of the legislature. The possibility that some local jurisdictions may adopt the model ordinance does exist regardless of the Legislature's decision.

At this point, only Palo Alto, California, has an ordinance similar to the model ordinance. While the model may not necessarily have an immediate impact on systems for hazardous gases, it has generated a significant amount of controversy. The Semiconductor Industry Association (SIA), the California Manufacturers Association (CMA) and the Compressed Gas Association (CGA) have all taken position in opposition to the model ordinance. Attempts are being made to reach a compromise.

2. Uniform Fire Code is a model developed by the Western Fire Chief's Association and published by the International Conference of Building Officials along with the Uniform Building Code, Electrical Code and others. These models may be subsequently adopted by local governing agencies to become local law.

   The Uniform Fire Code is issued every three years. Certain sections are reviewed, and each year a supplement issued. The current revision of Article 80 is of significant importance at this time. It has been adopted by the Western Fire Chief's Association, and implementation at the local levels should begin soon.

   One of the basic differences between the AB 1021 Report and Article 80 is that the Report covers toxic gases only; however, it does take a liberal interpretation of what it calls toxics. Article 80 covers all hazardous gases. The likelihood that it will have a near term impact is far greater. One significant fact is that gas cabinets are recognized specifically as an enclosure for storing and dispensing hazardous gases. Some of the requirements for gas cabinets are shown:

"DIVISION III STORAGE REQUIREMENTS...

80.303.a.1 Fire Extinguishing Systems...

A. Gas cabinets and exhaust enclosures for the storage of cylinders shall be internally sprinklered...

80.303.a.5 Exhaust ventilation.

A. Ventilated area. Storage of cylinders shall be within ventilated gas cabinets....Exhaust systems for gas cabinets, exhausted enclosures and separate gas storage rooms shall be designed to handle the accidental release of gas.

B. Gas Cabinets. When gas cabinets are provided, they shall be;

   i. Operated at negative pressure in relation to the surrounding area.

   ii. Provided with self-closing limited access ports or noncombustible windows to give access to equipment controls. The average velocity of ventilation at the face of access ports or access ports or windows shall be not less than 200 feet per minute (fpm) with a minimum of 150 fpm at any point of the access port or window.

iii. Connected to a treatment system.

iv. Provided with self-closing doors.

v. Constructed of not less than 12 gauge steel."

Other sections have specific requirements related to the piping system controlling the gas at the source:

"DIVISION IV DISPENSING, USE AND HANDLING...

80.401.c. Piping, valves, fittings and related components.

Piping, valves and fittings. Piping (including tubing), valves and fittings conveying hazardous materials shall be installed in accordance with approved standards and shall comply with the provisions of this subsection.

1. General. Piping, valves, fittings and related components used for hazardous materials shall comply with the following;
   A. Piping, valves, fittings and related components shall be designed and fabricated from materials compatible with the materials to be contained and shall be of adequate strength and durability to withstand the pressure, structural and seismic stress and exposure to which they may be subjected.
   B. Piping and tubing shall be identified in accordance with nationally recognized standards to indicate the material conveyed.
   C. Emergency shutoff valves shall be identified and the location shall be clearly visible and indicated by means of a sign.
   D. Back-flow prevention or check valves shall be provided when the back-flow of hazardous materials may create a hazardous condition or cause the unauthorized discharge of hazardous materials."

The code also has sections covering enunciators and toxic gas monitoring systems. There exists in legal requirements the elements of minimum safety systems for hazardous gas use.

B. Current status of systems for hazardous gas applications.
   1. Requirements for a good system.

      With the code defining the minimum requirements, what makes a good system for hazardous gas control?

      a. It obviously must meet code.
      b. It must be compatible with the gas and the process being used. Herein lies one of the major problems with some systems being manufactured. One can not design a totally standard system. It must be adaptable to specific gases. Furthermore, in many cases, specific processes have requirements that effect the system design. It is important that equipment designers know both the properties of the various gases and process operations. It is not enough to know who supplies components that can be fitted together.
      c. It is necessary that the systems be tested properly. Leak testing of individually fabricated subassemblies and completed modules (i.e., panel, manifolds) must be tested under pressure conditions that meet or exceed operating conditions. Inboard or vacuum type leak tests are not sufficient. Checking at one atmosphere pressure differen-

tial on a system that will operate at pressures of 2 to up to 150 atmospheres needs more than pencil and paper extrapolation.

A good system is one that first meets code requirements. Second, it is one that is designed and built by someone that knows and understands the material being used, the process being served and the working functions of the components used to construct the system.

2. Manual or Automatic.

If one considers only the functions being performed, there should be little difference between manual systems and automatic systems. Their goal is to expel unwanted gas, the mechanics are secondary. What is important is the reason for considering which system to use. If the selection is one of concern over the way in which personnel carries out the function, then there may be other problems. If the choice of an automatic system is made simply because it means less training of personnel, then the choice is improper. While it may be true that an automatic system will give more consistant reproducible purging, the operator should be qualified to perform the job manually. An automatic system, in addition to the reproducible nature of purging, should provide: first, the flexibility to allow the operator to perform other tasks while purging is taking place; second, a visual display of the operation; third, status information on the system at any given time; and fourth, a mechanism for total operation control and the means to communicate between a central control area and each source system.

3. Design variations.

There are many design variations among vendors, each must be carefully scrutinized. A well designed system will consider the following criteria:

Safety, Modularity and Serviceability.

Safety has already been considered. Modularity in this context means the ability to interchange components, to take one part out and replace it with another like or comparable part not necessarily by the same manufacturer. This requires that the physical design take into account the precise dimensions of each component and consider the availability of replacements.

There is another aspect to modularity that one can consider, that is the total purge module as a replaceable part. One of the newer designs available now uses this approach. The principle consideration is that if the system is so deteriorated that it requires a number of new parts, it is probably more cost effective to replace the entire unit than recondition it. The only parts in the basic module that are removable are the regulator and the emergency shutdown valve.

Serviceability relates to modularity. It is the ease with which any out-of-service system can be brought back into operation in a minimal time period.

C. Current status of components used in gas safety systems.

The selection of components requires vigilance because a number of changes are taking place with components, and selection of compatible components is very important.

1. CGA connections and cylinder valves.
   a. The CGA connection is probably the weakest component in the system. For more than forty years prior to 1985, there was little, if any, upgrade in both CGA connection design and cylinder valve design. A number of new designs are being studied and tested. In the case of CGA connections, which have always been vulnerable to mating surface damage, there are presently some new designs being studied by the Compressed Gas Association. The new designs involve face seal type connections with metal gaskets. The Compressed Gas Association is a very deliberate organization. They will recommend changes after both laboratory and field testing. Another problem area is the use of adapters to convert equipment designed for use with one gas to use with another, without rebuilding the equipment. This is not the proper way to solve the problem.
   b. Cylinder valves have at least two new innovative design changes in progress at this time. The first involves flow control from cylinder valves. Two different approaches to controlling flow out of a cylinder valve are being considered:
      1). A fixed orifice type of flow limiter has been designed and is being tested by the CGA. These results look favorable.
      2). An excess flow shutoff device as a part of the cylinder valve is also being studied. The results of this study should be available in the near future.

      The second innovative design change for cylinder valves involves the automating of the valve operation. Several designs of pneumatically operated valves are in test. Problems have surfaced in these tests which may slow down the widespread introduction of these valves.
2. Filters.

   The necessity to filter gases is well established. The placement of the filter(s) in the system remains a point of consideration. A filter of moderate particulate size should be placed as close as possible to the gas source. This protects the gas system from plugging. The use of filters smaller than 0.5 micron at the beginning of the gas system is not recommended because of flow restriction problems. A 0.02 micron filter at the point-of-use tool is recommended. Filters have a bad habit of loading with condensate in liquid system applications, such as chlorine and ammonia. If flow is high, the problem is increased. In these cases, it may be necessary to eliminate the filter at the source.
3. Tubing.

   Internal surface preparation of tubing is a debatable point. Most specifications require internal electropolish. The thought is, "the smoother the better - the less particulate generated." This may be true if no work is done on the tubing. However, there is significant evidence to show that mechanical work on electropolished surfaces may cause more particulate than it eliminates. A recent publication, February 1987, of Solid State Technology addresses the particulate from electropolished versus chempolished surfaces. Work done at the University of Texas several years ago also addressed this question. The problem is a subtle one. It is generally thought that a bright, shiny, mirror-like smooth finish is a pure metal surface. Not so; the metal roughness factor is reduced significantly, but the surface is covered with an oxide film. It is poor surface adhesion that produces the particulate. The problem with an electropolished surface is the rate and mechanism of the oxide film formation. The rate of formation in this case is much faster, thus less uniform and less adherent. A film formed after chemical polishing occurs at a slower rate, is more uniform, and is more adherent. What does this all mean? If little or no work is done on electropolished tubing, then the surface is not disturbed and particulate will not be generated. If it is necessary to make bends, there are two choices. Make the bends and accept that fine particles will be possible. Or make the bends, weld the tubing and then chempolish at completion. In doing this, a new problem is created. There is great difficulty in removing all of the acid and salts from the internal parts of the system. With either choice, the potential exists for the system to be slightly contaminated.

4. Fittings.
   Metal-to-metal face seal type fittings should be used. Since the original patent ran out, there have been a number of new entrants making these fittings. Most are now interchangeable with one another. The greater availability of fittings has been a benefit.
5. Regulators.
   Regulators are badly misused and abused. While they may be considered as a special kind of valve, regulators should not be used as a shutoff device except in emergency situations. The purpose for a regulator is to control the pressure of a gas being delivered to a point-of-use tool. The regulator will accept high inlet pressure of a cylinder gas and allow the user to reduce that pressure to a lower pressure for use. It should be noted that regulators are designed for specific applications and should, in most cases, be chosen especially for the application. One type is called a general purpose or standard regulator. This is misleading; the general purpose regulator is one whose applications are broader than others. Some misinformation has been circulated about regulator selections and needs to be corrected. A two-stage regulator is frequently recommended for silane and silane mixtures. This is a poor choice. If two degrees of regulation are desired, a far better choice is to use two single-stage regulators in

series. Install one in the source purge control module and one at the point-of-use tool. Two-stage regulators are very difficult to purge in the mid-range or center section. This means that the trapped air which remains can react with silane to produce SiO_2 (silica). The silica accumulates on the surface of the poppet valve seating area and prevents closure. A better choice is the two single-stage regulator design with a finite distance between them so purging can be complete. When an internal burn of silane occurs, the poppet valve is fouled and regulator creep is the result. All regulators operate basically the same way even though there are significant differences in physical design. To adjust or operate the regulator, one should follow a procedure such as:

a. Before connecting, turn handle counter clockwise to stop.
b. Connect regulator and open cylinder valve.
c. Turn the regulator handle slowly clockwise, watching delivery pressure gauge. Pressure will rise. Adjust to desired pressure.

The following takes place within the system, respectively:

a. Turn the handle counterclockwise, release all mechanical force from the bonnet spring. Both gauges should read zero.
b. Open cylinder valve, cylinder pressure will show on high pressure gauge and full cylinder pressure will be exerted on base of poppet valve.
c. Turn handle slowly clockwise, drive screw exerts force on diaphragm drive plate. Diaphragm is depressed against needle extension of poppet valve forcing it out of the seating area. Gas flows through the open valve until gas pressure under the diaphragm equals the spring pressure.

For most hazardous gas applications where the cylinder pressure exceeds 150 psig, the best regulator to use is a tied diaphragm or captured poppet type. These can be obtained in both single-stage and two-stage types. It has already been suggested that two-stage regulators are not the best selection for silane; the same can be said for other hazardous gases. A tied diaphragm regulator is one in which the tip of the poppet valve extends through the nozzle assembly and actually attaches to the bottom support plate of the diaphragm. The benefit of this design is evidenced in the silane case above. If improper purging has taken place and air remains inside the regulator, a burn will occur. SiO_2 on the seating area will make it difficult for the poppet valve to close. As gas pressure builds up under the diaphragm, it will pull harder on the poppet, thus giving the effect of trying to heal itself. Examples of some gases that tied diaphragm regulators should be used with are silane, silane mixtures, mixtures of arsine, phosphine, diborane, hydrogen, hydrogen chloride and ammonia. A caution should be noted concerning the compatibility of seat materials. A number of plastic materials are used as regulator seats. The literature indicates that Kalrez seats should be used with HCl and ammonia. Practical experience shows that

Kalrez frequently fails with both of these materials. Either Kel F or Tefzel are better choices.

Special single-stage regulators should be used for low pressure gases or liquid vapors being delivered to reduced pressure tools. These are known as absolute pressure regulators, vacuum regulators or reduced pressure regulators. In general, they are characterized by larger diaphragm areas, smaller orifices and more delicate spring action. If a standard regulator or a tied diaphragm regulator is used in low pressure applications, one may observe significant pressure swings on the delivery gauge. These swings will have a profound effect on a mass flow controller. Due to internal design of a standard regulator, condensation of the gas occurs inside the cavity with subsequent re-evaportion. These continual condensation and evaporation actions disrupt the gas flow at very low pressure differentials. The absolute pressure regulator generally corrects this problem.

6. Valves.
   The seat design of a valve is critical for longevity. Valve seats are usually made up of two parts, a polished metal surface and a plastic material. Some seats have a flat metal seating surface that provides a large contact area for the mating surfaces. Others have a smooth radiused metal side, while still others have a fairly sharp, almost knife edged metal side. The latter style will give good initial closure, but take a permanent set which requires more frequent adjustment or replacement of the seat. Several seat materials in common use are Kel F, Kalrez, Teflon, Tefzel, Viton and Buna N. Viton and Buna N are generally not applicable to seating material uses being considered. These may be used for gaskets or O-ring materials. In general, Teflon is not used as frequently as the others because Teflon tends to cold flow. Care must be exercised in the selection of seat material because experience in these applications indicates some acceptable compatibilities fail. An example is chlorine; generally Kel F, Teflon and Tefzel are considered compatible with it. In fact, Kel F will get soft and spongy in a chlorine medium. It loses its seating quality very rapidly. In this case, Tefzel and Teflon are better.
7. Specialty components for safety.
   a. Venturi vacuum generator, a name given to a gas aspirator. In operation, a gas stream is passed across an opening at a fairly high velocity. As the gas stream flows over this opening, molecules are carried out of the opening thus creating a vacuum. These units perform well; vacuums of 22 to 24 inches have been measured at the end of a 40 foot length of 1/4 inch OD tubing. The benefit of using the venturi vacuum generator in a system is the greater ability to remove undesirable gas during purging. If a flow limiter is used in the system, the time interval for vacuum venting may need to be increased.
   b. Excess flow valves are devices with a tainted history. The

early ones all leaked badly and a solution to this problem was to put a face seal blind nut cap on them, trapping gas in an inner area. When the nut was removed, out came the gas. Fortunately, the need for these devices stirred enough interest that now there are several available that do not leak and provide excellent service.

c. Excess flow sensor switches are in somewhat the same situation as excess flow valves. The most popular type provides an all welded construction with no leak path to the outside world. The use of an excess flow switch allows one to eliminate the excess flow valve unless safety redundancy is desired.
d. Pressure sensor switches and transmitters may be used in the system on both the high and low pressure sides. The sensor switch will indicate a "go/no go" condition, while the transmitter puts out a signal that can be read directly as pressure. The latter is of particular importance if remote status checks are required or for data storage.
e. Exhaust flow sensor switch signals the loss of exhaust flow and serve either as an alarm or emergency shutdown signal.
f. Seismic sensor switch or valve is perhaps the least valuable of the sensors currently being used in some systems. The AB 1021 Report, as it reads now, requires a seismic device. It is a cause device, that is, it signals a seismic event has occurred. It does not necessarily mean a problem exists.
g. Toxic leak detectors are required by both AB 1021 and Article 80 for toxics. They are an absolute must for safe operation. A number of detectors are on the market and much consideration should be used in selecting one.
h. Low cylinder gas pressure/weight sensor will signal when it is time to change the process gas cylinder.

A host of other sensing devices may be considered for specific applications.

Complete gas management systems may be incorporated if cost is not a significant component of the decision process. With the large number of vendors in the gas cabinet and purge control module business, great care must be exercised in selecting the proper supplier.

PREVENTING AND CONTROLLING ACCIDENTAL GAS RELEASES

P.D. Moskowitz, V.M. Fthenakis, and P.D. Kalb
Biomedical and Environmental Assessment Division
Brookhaven National Laboratory
Upton, New York 11973

ABSTRACT

Toxic, flammable, and explosive gases may be used in photovoltaic cell research laboratories and in commercial manufacturing facilities. Accidental release of these materials can present hazards to life and property. Accidents can arise from a variety of mechanical and human related failures. These can occur from the time materials are received at the loading dock of the facility to the time treated gases are discharged to the atmosphere through a stack. Each type of initiating event may require a different control approach. These may range from the training and certification of plant workers charged with the handling of gas cylinder hookups to installation of emergency pollution control systems. Since engineering options for controlling released materials are limited, emphasis should be placed on administrative and engineering approaches for preventing such accidents. These are likely to be the most effective approaches for protecting life and property.

INTRODUCTION

A diverse set of toxic, flammable, and explosive gases are used in photovoltaic cell production. Accidental release of these materials can present significant hazards to life and property. As a result, strategies for preventing and controlling accidents during the storage, use, and disposal of these gases are needed. In response to these needs, this paper reviews approaches for identifying hazardous gases at photovoltaic cell facilities that require the establishment of prevention and control strategies and types of incidents which can result in an accidental release. Also, new regulatory controls (federal, state, and local), which impact facility operations, and hazard management options for preventing or controlling incidents are discussed.

HAZARD IDENTIFICATION

The first step to reduce risks to life and property from toxic or hazardous gas releases is to identify and evaluate those chemicals or activities that require the establishment of prevention and control strategies. Performance of these identification and ranking analyses may range from simple to elaborate data collection and evaluation efforts.

Often, environmental auditing is the first step in this overall process.[1] An audit is a systematic, documented, periodic, and objective review of facility plans, operations, and practices related to environmental issues. Audits can be designed to accomplish any or all of the following: verify compliance with environmental requirements; evaluate the effectiveness of environmental management systems; or assess risks from hazardous materials, equipment, and practices. Usually, a number of steps is involved. They include: audit planning, on-site field audit, audit evaluation, reporting, and post-audit follow-up.

An environmental audit begins with a meeting of audit and facility management teams to gather and discuss facility information (organization, process layout), environmental management systems, regulatory requirements, and corporate policies. Collection and review of this advance information will result in a more detailed audit plan outlining the needed audit steps, how each is to be accomplished, who will do them, and in what sequence. The prime purpose of this important first step is to develop a working relationship between these teams, and to understand facility processes, internal controls (both management and engineering), plant organization and responsibilities, compliance parameters, other applicable requirements, and any special problems.

After the initial audit planning is completed, a formal on-site field audit is conducted. Through inquiry, observation, and verification, the following issues are examined:

(1) What types of equipment are used?
(2) What types and quantities of materials are needed in these processes?
(3) How are these materials stored and handled?
(4) What types and quantities of waste are produced?
(5) What types of occupational and environmental controls and safety systems are incorporated into the production processes?
(6) How do workers interact with the process equipment?

After the on-site field audit is completed, analyses are prepared to evaluate the soundness of the facility's environmental management systems. At the culmination of this effort, a written audit report is prepared. The purpose of this report is to provide information about potential hazards, potential compliance-related issues, and corrective actions.

As a follow-up and adjunct to auditing, screening tools are often used to identify those activities or processes which present the greatest hazard and require the most attention. Industrial,[2] national,[3] and international[4,5] organizations have published guidelines for these purposes.

Of these, the interim U.S. Environmental Protection Agency (EPA) guidelines[3] have attracted great interest. They include two sections of particular interest. First, is a list of "acutely toxic chemicals." Chemicals were included in this list if their median lethal dose (LD_{50}) or median lethal concentration (LC_{50}) in mammalian test species was less than 50 mg/kg, 25 mg/kg, or 0.5 mg/l for dermal, oral, or inhalation routes of exposure, respectively. Chemicals used by the photovoltaics industry which appear in the EPA list are shown in Table I. The second part of these guidelines includes a quantitative tool to determine the quantity of a chemical which can present an acute risk to public health. These estimates can be used to rank these chemicals. The basis for these estimates is information on the toxicity of the specific chemical and its dispersive potential in the atmosphere. In this calculation, the EPA recommends that the OSHA Immediately Dangerous to Life and Health (IDLH) concentration[6] be used as the toxicity index. The IDLH represents a maximum level from which one could escape within 30 minutes without any escape-impairing symptoms or any irreversible health effects. In the guidelines, atmospheric dispersion potential is estimated by implementing a nomogram built using simple dispersion assumptions (e.g., neutral buoyant, nonreactive plume) with external (i.e., distance to plant boundary) inputs. Application of this approach produces an estimate of the threshold quantity of a specific compound which can present an acute hazard to public health in the event of a catastrophic release of the material stored on-site. By comparing the estimated threshold quantities with the on-site inventories, the relative hazard among different materials can be ranked. The IDLH values and estimated threshold quantities at three distances for the "acutely toxic compounds" used by the photovoltaics industry are also shown in Table I.

A second approach is to compare the on-site inventories with the "Reportable Quantities" (RQ) defined by the EPA under authorities of the Comprehensive Environmental Response, Compensation and Liability Act of 1980 (CERCLA).[7] These represent quantities of materials, which when released to the environment in an accident, must be immediately reported by facility managers to the National Response Center. In these comparisons, it can be simply assumed that the larger the ratio of the on-site inventory and the RQ, the larger the hazard. The RQ's for some materials used in photovoltaic cell manufacture are shown in Table II.

TYPES OF ACCIDENT INITIATING EVENTS

In the storage, handling, use, and disposal of toxic gases, accidental releases may arise for any of the following reasons: design errors, construction errors, equipment failures, human errors, and process upsets. Marsh and McLennan Protection Consultants[8] have analyzed 100 major incidents (fire, explosion, and toxic release) in the hydrocarbon-chemical industry and classified the causes of these events into the following categories: mechanical equipment failure, process upsets, human error, and arson or sabotage. Figure 1 shows their compiled statistics on these incidents.

Component failure modes and rates for equipment used in the nuclear industry have been studied extensively.[9] Failure of individual components in isolation may not, however, result in the accidental release of a toxic gas. Often, a release may occur only after a series of components fail. Various techniques exist to examine both the probability and relationship between component failure and accident occurrence: preliminary hazard analysis, failure mode and effects analysis, failure mode effect and criticality analysis, parts count method, fault hazard analysis, fault tree analysis, event tree analysis, simulation, double failure matrix, cause-consequence diagrams, relative ranking, epidemiological approach, and reliability analysis. The strengths and weaknesses of these approaches are described in detail elsewhere.[2,10] A brief synopsis of two commonly used approaches is presented below.

Failure mode and effect analysis (FMEA) is used to analyze all failure modes of a given item of equipment for their effects on other components and the overall system. This method is equipment oriented and is a bottom-up type of approach. Actual steps in FMEA include: identify and list all components, identify all failure modes, determine effects of each failure on other components and the overall system, calculate the probability of each failure mode, and evaluate the importance of each failure mode.

Fault tree analysis (FTA) as contrasted with FMEA is a deductive or top-down approach. Construction of the fault tree is the most critical step in this type of analysis. Construction begins by defining an undesirable event (e.g., accidental release of a solvent to a groundwater system) and repeatedly asking how that might come about, until the basic causes or lowest practical level of detail is reached. FTA can also be used to estimate quantitatively the probability of an accident or accident initiating event. This requires examination and application of reliability data for components.

Although these techniques have enjoyed widespread usage in the aviation and nuclear industries, they have had only limited application in the chemical industry and in only one published case to the photovoltaics industry.[11] Consequently, there are only limited examples to draw from and more importantly only limited data have been compiled of direct relevance to the photovoltaics industry.

REGULATORY CONTROLS

Growing awareness of the risks involved in the use of highly toxic chemicals for manufacturing has led to a dynamic regulatory climate with increasingly stringent control and safety procedures imposed by federal, state, and local authorities. Presented below are summaries of some of the significant regulatory programs that have been adopted or are being discussed by various levels of government.

The Comprehensive Environmental Response, Compensation, and Liability Act of 1980 (CERCLA, also known as the Superfund Act) was enacted by Congress to respond to releases of hazardous substances to the environment either through accidental spills or chronic releases, e.g., malfunctioning or abandoned dump sites. By mandating authority over the cleanup of hazardous waste contamination, CERCLA compliments earlier legislation including the Federal Water Pollution Control Act (Clean Water Act), the Resource Conservation and Recovery Act (RCRA), the Clean Air Act and the Toxic Substances Control Act (TSCA) in providing "cradle to grave" oversight of materials which potentially threaten human health and safety and the environment. CERCLA requires notification to the National Emergency Response Center immediately upon release of hazardous materials in excess of "CERCLA reportable quantities" (see Table II).

The Superfund Amendments and Reauthorization Act of 1986 (SARA) re-emphasized the federal effort to reduce risks to health and safety and the environment from exposure to toxic materials by: i) increasing available funds for waste site cleanup; ii) revising many sections of the original law (CERCLA) in favor of stricter standards, schedules, and enforcement powers; and iii) adding several new sections that expand the scope of the original law. In response to the catastrophe at Bhopal, India, and other recent chemical emergencies, one of the new sections adopted is entitled the "Emergency Planning and Community Right-to-Know Act of 1986," under Title III of SARA. The provisions included in this section require disclosure of information on the dangers of hazardous materials to workers and the public.

A major focus of Title III is to further involve state and local authorities in planning and response functions for chemical emergencies. In this regard, Section 301 mandates the formation of state emergency response commissions and local emergency planning committees to take responsibility for handling chemical mishaps. These agencies must prepare comprehensive response plans, review data from qualifying facilities within their jurisdiction, and designate community and facility emergency response coordinators.

Title III mandates that releases of CERCLA hazardous substances in excess of reportable quantities now require notification of local emergency planning authorities as well as the National Response Center. In addition, EPA has established a list of 402 "extremely hazardous substances" along with both "threshold planning quantities" and "reportable quantities" of these materials.[12] Immediately following accidental releases of "extremely hazardous materials" in excess of "reportable quantity" levels (see Table II), emergency notification must be given to the community emergency coordinator and to the state emergency planning commission. Emergency notification must include: the chemical name/identity, estimated quantity of release, time and duration of release, media into which release occurred, known or anticipated health risks, and recommended precautions to prevent or minimize risks. Follow-up notification concerning actions taken and relevant health data is also required. If a hazardous material is present at a facility above "threshold planning quantities" (See Table II), additional notification procedures are called for. Title III requires qualifying facilities to provide material safety data sheets and annual emergency and hazardous chemical inventory forms to state and local authorities. The latter are to include an estimate of the maximum amount of each hazardous chemical present at any time during the previous year, an estimate of the average daily inventories, and locations of the materials. The final reporting requirement of Title III (under Section 313) requires certain facilities to prepare annual toxic chemical release forms that detail the chemical usage, waste treatment/disposal methods, and estimate the annual quantities of chemicals released. These requirements apply to owners and operators of facilities that have 10 or more full-time employees, are in Standard Industrial Classification Codes 20-39, and use 10,000 pounds per year or manufacture/process 75,000 pounds per year of a toxic chemical at the facility.

The State of New Jersey has recently passed a law known as the Toxic Catastrophe Prevention Act (State of New Jersey 1986). Under this law, facilities that generate, store, or handle substances listed on the "Extraordinarily Hazardous Substances List" (EHSL) established by the state's Department of Environmental Protection (DEP) are required to register with the DEP. Such registered facilities are required to submit either a "Risk Management Program" or a "Extraordinarily Hazardous Substance Accident Risk Assessment" that is acceptable to the DEP. Although the law specifies some general requirements for these programs, specific criteria and standards to evaluate such programs have not yet been promulgated.

Interest in managing risks to health and safety at the local level is exemplified by the recently proposed Santa Clara County, California, Toxic Gas Model Ordinance. Motivated by the density of high technology manufacturing facilities located in the area commonly known as "Silicon Valley," the local fire chief's association prepared a comprehensive set of regulations designed to minimize hazards. A broad range of planning and emergency response provisions are incorporated within the ordinance including extensive design and construction standards, process and safety equipment performance standards, training and safety protocols, monitoring specifications, and dispersion modeling requirements. Reporting requirements that overlap and in some areas go beyond those required under SARA Title III, include a Hazardous Material Management Plan, Hazardous Materials Inventory Statement, and plans for emergency response to a release or threatened release of toxic gases. All local facilities that use toxic gases (as defined by the EPA list of Extremely Hazardous Substances or several other sources), without regard to specific quantities, are subject to this ordinance unless specifically exempted by the local Fire Chief. Exemptions are granted where it can be demonstrated that based on quantity, toxicity or concentrations, the material does not "present a significant actual or potential hazard to the public health, safety, or welfare."

Construction of photovoltaic research and production facilities are subject to existing building and fire codes that establish minimum standards for safe practices in order to protect life and property. Since codes are generally enforced at the most local level of government (i.e., city or town) and are subject to some variation, model codes have been developed to provide guidance to municipalities and technical consistency among codes. The Uniform Building Code and Uniform Fire Code are widely accepted examples of these model codes. In recognition of the hazards associated with some of the processes used in the semiconductor industry, new provisions were incorporated into the Uniform Building Code and the Uniform Fire Code by the respective official fire and building code bodies. The provisions of these codes applicable to the semiconductor industry (and by extension the photovoltaics industry) are included in the H-6 Design Guide to the Uniform Codes for High Tech Facilities. Detailed design standards cover areas such as building materials, construction methods, layout (including maximum area allowed for hazardous production material storage), equipment, and safety specifications. Procedural issues such as storage and handling practices (including maximum storage quantities for hazardous materials and compatibility constraints for reactive materials), spill prevention, monitoring, labeling/warning signs, and emergency drills are also covered in the code requirements.[13]

MANAGEMENT OPTIONS

Toxic gas releases can arise from a variety of human and mechanical related failures. These can occur from the time materials are received at the loading dock of a facility, to the time treated gases are discharged to the atmosphere through a stack. Each type of initiating event may require a different control approach. The specific options will depend on the materials, their applications, and the nature and quantities of material which can be accidentally released. A material by material analysis of such options is beyond the scope of this report; some material-by-material analyses, however, are presented elsewhere.[11,14,15] Consequently, only general guidelines are presented below. These are divided into administrative and engineering controls.

Administrative

Administrative approaches may include but not be limited to:

(i) Establishment of employee training and certification programs to handle normal operating conditions and upsets and emergencies. The programs should include written materials, class-room type instruction, field drills, and where necessary testing and certification standards.

(ii) Implementation of housekeeping and maintenance programs including regular inspection of toxic gas handling and delivery systems.

(iii) Establishment of formal process review and certification teams for activation of new or modified equipment and materials.

(iv) Isolation of workers in time and space from operations which may generate contaminants in such quantities that could present hazards to health. Enclosing dangerous operations together in separate rooms or buildings not only sharply reduces the number of workers exposed but may also simplify control procedures. An opposite approach to this is to isolate the personnel in closed ventilated work spaces under positive pressure to reduce worker exposures to contaminants.

(v) Establishment of emergency response teams.

(vi) Establishment of decontamination and emergency procedures including but not be limited to such activities as liaison with outside authorities, appointment and definition of the roles and responsibilities of emergency response personnel, development of site action plans including fire-fighting procedures and rescue systems, and plant shut-down procedures. After the situation is brought under control, procedures to decontaminate work areas, equipment, and clean-up workers should be implemented.

(vii) Testing of emergency response plans and procedures to anticipated emergencies including loss of power, fire, and explosion.

Engineering

In the area of engineering options, the first and perhaps most important factor in controlling toxic gases is to incorporate inherently safer features into the facility's original design. EPA in its recent manual on controlling accidental gas releases[16] divides these options into the following areas: siting and layout, structures and foundations, vessels, piping and valves, process machinery, and instrumentation. Examples of each issue area are given in Table III.

Although preventing accidental releases through the choice of inherently safer technologies, processes, systems and safe procedures is of the utmost importance, vapor/gas releases may happen, in spite of all precautions. Therefore, techniques to mitigate these hazards should be in place. One of the difficulties encountered when responding to a hazardous vapor discharge is the relatively short period of time between the release and subsequent dispersion of the vapors. However, denser air gases (e.g., hydrogen selenide) and vapors generated from liquid spills (e.g., metalorganics) disperse relatively slowly under certain meteorological conditions (e.g., several minutes to disperse to distance of 100 ft under calm conditions) and may allow some time for response.

Control techniques for unconfined or partially confined releases include: inactivating the gas by burning it; entraining it by secondary confinement or foam spraying; dispersing it by blowing, heating, or water spraying; and entraining and inactivating it by chemically reactive foams or liquid jets. Burning of a gas may be applicable only to certain flammable gases. Heating of the gas (e.g. by propane burners around the perimeter of a plant) may result in the enhanced dilution of a heavy gas cloud. Secondary confinement can be very effective but is very expensive and can be applied only to stationary systems. Furthermore, these systems may also be susceptible to fire and explosions. Also, they may be susceptibile to failure from the same events that caused the initial leakage (e.g., earthquake, fire). Foam spraying is efficient in suppressing vaporization of a liquid spill by covering it. Reactive foam spraying can be effective if the contaminated air is sucked inside the foam generating machine. This, however, is difficult to do in an unconfined release. Water spraying (i.e., water curtains) is the only mitigation technique with data on its effectiveness existing in the literature. It has been used with limited success in several experimental tests of unconfined releases and in real emergencies. The dilution effect of this technique is, in general, on the order of 10 which is not sufficient for highly toxic gases which are hazardous to very low concentration levels; they may require an order of magnitude or higher dilution. However, in at least two cases of controlling water soluble gases, the effectiveness of this control was high. In both cases spraying was done in such a way as to trap and absorb the gas within the liquid instead of dispersing it. In a series of large-scale field tests where water from fire hoses was sprayed on ammonia,[17] the technique was estimated to be up to 99.8% efficient. In another series of field tests where water was sprayed on hydrogen fluoride, the effectiveness was reported to be approximately 36 to 49%.[18]

CONCLUSIONS AND RECOMMENDATIONS

Both adminstrative and engineering options are available to the photovoltaics industry to prevent and control toxic gas releases. Since engineering options for controlling released materials are limited, emphasis should be placed on administrative and engineering approaches for preventing such accidents. These are likely to be the most effective approaches for protecting life and property.

REFERENCES

1. U.S. Environmental Protection Agency, Environmental Auditing Policy Statement, Federal Register 50(217):46504-46508 (1985).

2. Goldthwaite, W.H.; P. Baybutt; F.L. Leverenz; and H.S. Kemp, Guidelines for Hazard Evaluation Procedures: Preparation and Content, (American Institute of Chemical Engineers, NY, 1987).

3. U.S. Environmental Protection Agency, Chemical Emergency Preparedness Program - Interim Guidance, (Washington, DC, 1985).

4. United Nations - International Labour Office, Working Paper on Control of Major Hazards in Industry and Prevention of Major Accidents, Rep. ILO MHC/1985/1, (Geneva, Switzerland, 1985).

5. World Bank, Manual for Industrial Hazard Assessment Techniques, Office of Environment and Scientific Affairs, (World Bank, Washington, DC, 1985).

6. Mackison, F.W. and R.S. Stricoff, Editors, NIOSH/OSHA Pocket Guide to Chemical Hazards, DHEW (NIOSH) Publication No. 78-210 (1978).

7. U.S. Environmental Protection Agency, Notification Requirements; Reportable Quantity Adjustments, Federal Register 50(65):13456-13522 (1985).

8. Marsh and McLennan Protection Consultants, One-Hundred Losses, A Thirty-Year Review of Property Damage Losses in the Hydrocarbon-Chemical Industries, Ninth Edition, (Chicago, IL, 1986).

9. Institute of Electrical and Electronics Engineers, IEEE Guide to the Collection and Presentation of Electrical, Electronic, Sensing, Component and Mechanical Equipment Reliability Data for Nuclear Power Generation Stations, (IEEE, NY, 1977).

10. Ozog, H. and L.M. Benedixen, Hazard Identification and Quantification Chemical Engineering Progress 83(4):55-64 (1987).

11. Moskowitz, P.D.; V.M. Fthenakis; R.W. Youngblood; and S.R. Mendez, Evaluating Risks Associated with the Use of Silane, Presented at the Hazard Assessment & Control Technology in Semiconductor Manufacturing, (Cincinnati, OH, October 21-22, 1987) .

12. U.S. Environmental Protection Agency, Emergency Planning and Community Right to Know Programs, Federal Register 51(221):41570-41591 (1986).

13. Goldberg, A. and L. Fleur, H-6 Design Guide to the Uniform Codes for High Tech Facilities, GRDA Publications, (Mill Valley, CA, 1986).

14. Fthenakis, V.M and P.D. Moskowitz, Characterization and Control of Phosphine Hazards in Photovoltaic Cell Manufacture, Solar Cells (in press).

15. Lee, J.C. and P.D. Moskowitz, Hazard Characterization and Management of Arsine and Gallium Arsenide in Large-Scale Production of Gallium Arsenide

Thin-Film Photovoltaic Cells, (BNL 36643, Upton, NY, 1985); Solar Cells 18:41-54 (1987).

16. Davis, D.S., G.B. DeWolf; and J.D. Quass, Prevention Reference Manual: User's Guide Overview for Controlling Accidental Releases of Air Toxics, (EPA/600/9-87/028, Research Triangle Park, NC, 1987).

17. Greiner, M.L., Emergency Response Procedures for Anhydrous Ammonia Vapor Releases, Plant/Operations Progress 3(2) (1984).

18. Blewitt, D.N., et al., Effectiveness of Water Sprays on Mitigating Anhydrous Hydrofluoric Acid Releases, pp. 155-171, in J. Woodward, Editor, Proceedings of the International Conference on Vapor Cloud Modeling, (American Institute of Chemical Engineers, NY, 1987).

Table I. EPA Acutely Toxic Chemicals and Estimated Threshold Quantities

Material	IDLH (g/m^3)	Threshold quantities (lbs) 100 m	200 m	500 m
Arsine [AsH_3]	0.024	~ 1.5	6	30
Boron Trifluoride [BF_3]	0.3	20	~ 70	370
Cobalt (fume and dust) [Co]	0.02	~ 1.5	5	22
Diborane [B_2H_6]	0.04	~ 2.8	9	47
Fluorides	0.5	32	110	600
Hydrogen fluoride [HF]	0.013	~ 1	3	17
Hydrogen selenide [H_2Se]	0.008	< 1	~ 2	9
Nitric acid [H_3NO_3]	0.250	~ 17	~ 60	300
Nitric oxide [NO]	0.12	9	30	150
Nitric dioxide [N_2O_4]	0.09	6	20	100
Ozone [O_3]	0.02	1.5	5	22
Phenol	0.38	25	85	470
Phosphine [PH_3]	0.267	20	60	300
Sulfuric acid [H_2SO_4]	0.08	5	~ 18	100
Toluene 2,4 diisocyanate	0.07	~ 4.5	~ 16	80

Table II. EPA List of Extremely Hazardous Substances, Threshold Planning Quantities, and Reportable Quantities (RQ) for Chemicals used in the Semiconductor and Photovoltaics Industry

Hazardous Substance	SARA Threshold Planning Quantity, lbs[a,b]	SARA Title III Reportable Quantities, lbs[a]	CERCLA Reportable Quantities, lbs[a]
Acetic acid	-	-	5000
Acetone	-	-	5000
Ammonia	500	100	100
Ammonium bifluoride	-	-	100
Ammonium hydroxide	-	-	1000
Aniline	1000	5000	5000
Antimony[c]	-	-	5000
Arsenic[c]	-	-	1
Arsenic trioxide	100/10000	5000	5000
Arsenic trichloride	500	5000	5000
Arsine	100	1	1
Boron trichloride	500	1	1
Boron trifluoride	500	1	1
Cadmium[c]	-	-	1
Cadmium oxide	100/10000	1	1
Carbon tetrachloride	-	-	5000
Chlorine	100	10	10
Chloroform	10000	5000	5000
Chromic acid	-	-	1000
Chromium[c]	-	-	1
Cobalt	10000	1	-
Copper[c]	-	-	5000
Diborane	100	1	1
Epichlorohydrin	1000	1000	1000
Fluorine	500	10	10
Formaldehyde	500	1000	1000
Gallium trichloride	500/10000	1	1
Hydrochloric acid	-	-	5000
Hydrofluoric acid	-	-	100
Hydrogen chloride (gas)	500	1	-
Hydrogen peroxide[d]	1000	1	1
Hydrogen selenide	10	1	1
Hydrogen sulfide	500	100	100
Lead[c]	-	-	1
Mercury	-	-	1
Methanol	-	-	5000
Methylene chloride	-	-	1000
Methyl ethyl ketone	-	-	5000
Nickel[c]	10000	1	1
Nitric acid	1000	1000	1000
Nitric oxide	100	10	10

Table II. (continued)

Hazardous Substance	SARA Threshold Planning Quantity, lbs[a,b]	SARA Title III Reportable Quantities, lbs[a]	CERCLA Reportable Quantities, lbs[a]
Nitrogen dioxide	100	10	10
Ozone	100	1	1
Phenol	500/10000	1000	100
Phosphine	500	100	100
Phosphorus oxychloride	500	1000	1000
Phosphorus pentoxide	10	1	1
Potassium cyanide	100	10	10
Potassium hydroxide	-	-	1000
Selenium[c]	-	-	1
Selenium dioxide	-	-	10
Silver[c]	-	-	1000
Sodium hydroxide	-	-	1000
Sodium hypochlorite	-	-	100
Sulfuric acid	1000	1000	1000
Tellurium	500/10000	1	-
Titanium tetrachloride	100	1	1
Toluene	-	-	1000
Trichloroethylene	-	-	1000
Vinyl acetate	1000	5000	5000
Xylene	-	-	1000
Zinc[c]	-	-	1000
Zinc phosphide	500	100	100

[a] Dash (-) indicates substance not subject to the notification requirements indicated.
[b] Extremely hazardous substances that are solids are subject to either of two threshold planning quantities as shown. The lower quantity applies only if the solid exists in powdered form and has a particle size less than 100 microns; or is handled in solution or in molten form; or meets the criteria for a National Fire Protection Association (NFPA) rating of 2, 3, or 4 for reactivity. If the solid does not meet any of these criteria, it is subject to the upper (10,000 lb) threshold planning quantity.
[c] Released particles < 100 microns.
[d] Concentration > 52%.

Table III. Factors to be Considered in Designing Inherently Safer Facilities

Siting and Layout

- Availability of reliable backup water and power supplies
- Minimization of effects of natural calamities such as earthquakes and fire through design features
- Segregation of hazardous activities into controlled areas
- Incorporation of fire/explosion barriers
- Establishment of easily accessible escape routes

Structures and Foundations

- Consideration of normal and abnormal load and vibration conditions
- Incorporation of equipment to prevent the accidental detachment of process lines conveying toxic chemicals

Vessels

- Overflow and vacuum protection
- Protection from damage caused by collision or vibration
- Implementation of testing, inspection, and maintenance procedures

Piping and Valves

- Avoidance of unnecessary dead-ends or rarely used pipe branches
- Use of welded stainless steel tubing
- Installation of remotely controlled shut-off valves
- Backflow protection

Process Machinery

- Installation of instrumentation to identify changes in pressure levels into and out of process equipment
- Backup power supplies for critical systems
- Surge protection for electronic controllers

Instrumentation

- Installation of fail-safe type equipment
- Redundancy of key components
- Use of trip systems for emergency situations

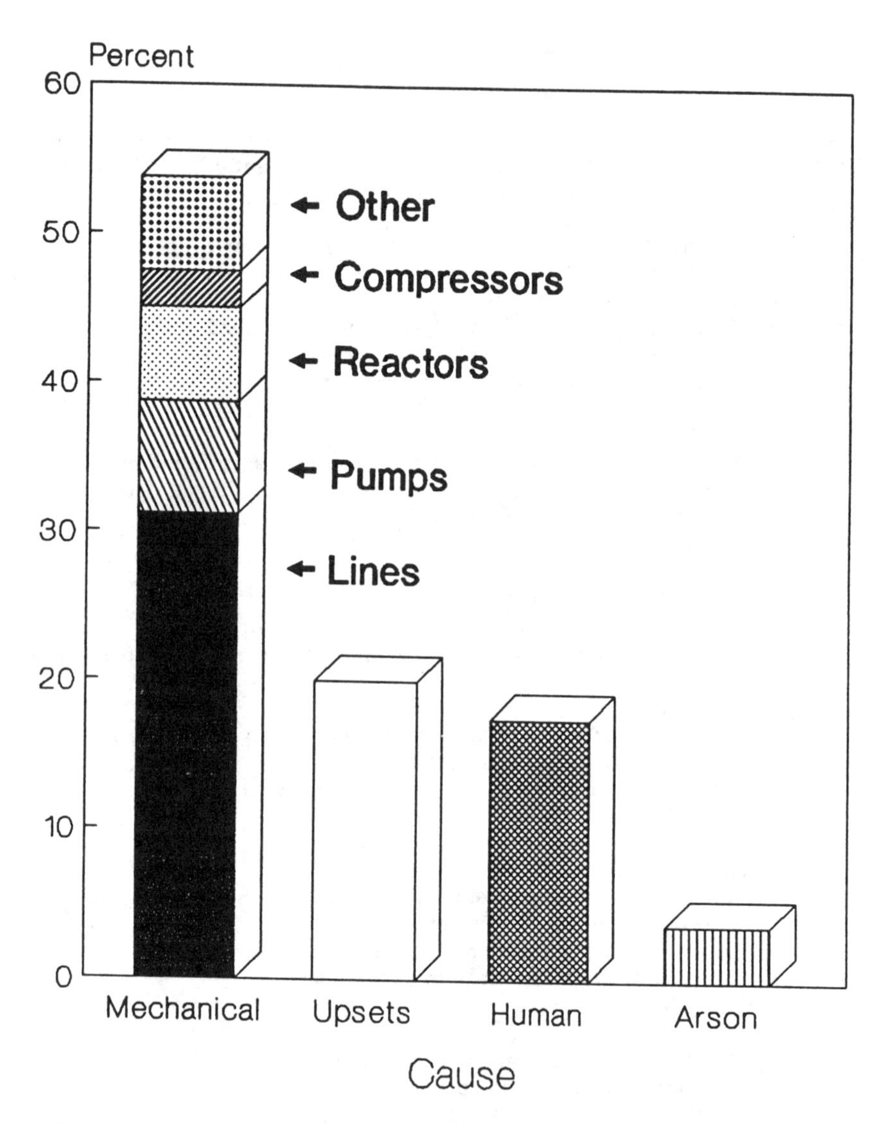

Adapted from Davis et al. 1987

SILANE-O_2 EXPLOSIONS, THEIR CHARACTERISTICS AND THEIR CONTROL

M.A. Ring, H.E. O'Neal and J. Famil-Ghiriha
Department of Chemistry, San Diego State University
San Diego, Calif. 92182

ABSTRACT

Prior results on the stoichiometry, upper pressure explosion limits and reaction mechanism of SiH_4-O_2 explosion reactions are discussed, and new data on the effects of added disilane on the pyrophoric and explosive characters of metastable SiH_4-O_2 mixtures are presented. The results have possible application to the prevention of serious explosions due to silane leaks into air.

INTRODUCTION

The highly explosive nature of SiH_4-O_2 mixtures has been recognized since the pioneering work of Freidel and Landenberg[1] in 1867; however, only recently has this problem received much attention. One of the major problems in the handling of silane is that when silane leaks into air an explosion or burning does not always immediately occur. Thus small leaks of silane have been known to produce sizable quantities of silane-air mixtures, which have subsequently detonated with tremendous violence. In an effort to better understand SiH_4-O_2 explosion reactions, we have made systematic investigations of their reaction stoichiometries as a function of initial mixture concentrations and have reexamined some aspects of their previously reported[2-4] upper pressure explosion limits (P_2). One interesting observation of this study[5] was that metastable SiH_4-O_2 mixtures could be prepared easily, and that their subsequent explosions (if any) were often spontaneous and unpredictable. This is consistent with the silane leak-explosion observations. In this paper, we review our previous results on SiH_4-O_2 explosions, and report recent results on the preparation and explosive characters of metastable SiH_4-O_2 mixtures. We also report on efforts to make silane-oxygen and silane-air mixtures more immediately pyrophoric through the additions of small amounts of disilane.

Prior Study Results[5]: Stoichiometric results for SiH_4-O_2 explosions, presented on a 10 mol reactant basis as a function of percent SiH_4 in the reactant mixture, are summarized in Table I. Silane consumption was quantitative in these explosions for all mixtures ranging in composition from 30 to 90% silane, and the only volatile products were H_2 and H_2O. Particularly significant is the volatile product composition as a function of percent SiH_4 in the explosion mixtures. Thus H_2O was the only volatile product of the 30% SiH_4 explosions, while H_2 was the only volatile product of explosions in mixtures containing 70% SiH_4 or more. Also shown in Table I are the calculated heats of reaction for two cases: 1) all products, including Si, SiO, and SiO_2, are gaseous, and 2) only H_2 and H_2O are gaseous.

Some interesting and mechanistically important inferences can be drawn from these data. First, branching chains of the explosion are propagated mainly, if not exclusively, by H atoms. Second, in oxygen rich mixtures, the $H_2 + O_2$ explosion contributes to the overall reaction. Third, part of the overall exothermicity of SiH_4-O_2 explosions comes from rapid condensation of the involatile products (i.e. oxides of silicon and silicon metal). Fourth, the induced thermal decomposition of silane to its elements is the main reaction in the explosions of silane rich mixtures. In the following we provide arguments to support these conclusions.

That the branching chain reactions of the explosions are propagated by H atoms follows from the fact that no water is observed in the products of explosions with silane contents of 70% and above. Thus if hydroxyl radicals were involved in chain propagation, water would surely be an observed reaction product. The only possible exception to this would be if water could act as an oxidizing agent toward silane under explosion conditions. That this is not the case follows from the observations that small quantities of water, added prior to the explosions of silane rich mixtures, survived the explosions intact. Hence chains of silane-oxygen explosions are propagated by H atoms, not OH radicals. This in turn bears on the second conclusion. Water is an important product of the explosions of oxygen rich mixtures. Since water is not a direct product of the SiH_4-O_2 branching chain reactions, it must arise from some other source. The most reasonable is the $H_2 + O_2$ explosion since the silane-oxygen explosion produces both the hydrogen and the high temperatures needed to initiate this explosion. The third conclusion follows from the reaction enthalpy calculations of Table I. Since an explosion reaction must be overall exothermic, condensation of the involatile products (Si, SiO, and SiO_2) is a necessary condition for explosions in mixtures with silane compositions exceeding 70%. It was found that explosions could be initiated in mixtures with silane contents as high as 94%! This requires extremely rapid condensation of the involatile products (i.e., condensation must occur in times appreciably shorter than explosion times) otherwise the reaction exothermicity would not be sufficient to maintain the explosion and drive it to completion. Finally, with regard to the fourth conclusion, it is clear from the reaction stoichiometries that oxidation reactions in silane rich systems represent a small fraction of the overall reaction. For example, in the explosion of 90% SiH_4-10% O_2 mixtures silane consumption was complete even though in the products, less than 11% of the silicon and <u>none</u> of the hydrogen was oxidized. Most of this reaction, therefore, occurred via the induced thermal decomposition of silane to its elements. Data from explosions of 90-94% silane mixtures support this conclusion. In these explosions, silane was not completely consumed, and disilane was produced in yields comparable to those observed in the pure silane pyrolysis (i.e., the steady state ratio ($[Si_2H_6]/[SiH_4]$ $\approx$ 0.03).[6] To produce and maintain a thermal explosion in oxygen poor-silane rich mixtures, adiabatic heat release from the silane oxidation must be able to generate reaction temperatures above the thermal explosion limit. This is not possible for mixtures with silane concentrations above 94% as these mixtures cannot be made to explode either by pressure reduction, or by temperature rise, or even by more extreme measures such as sparking with the discharge of a Tesla coil[5]. An explosion limit of around 94% silane is supported by calculations of reaction adiabatic flame temperatures. Thus for 90% SiH_4-10% O_2 mixtures, T(adiabatic) = 1560 K, where the estimated silane decomposition lifetime[7] is about 10^{-5} sec (i.e., short compared to the millisecond explosion durations observed); for 95% SiH_4-5% O_2 mixtures, T(adiabatic) = 1014 K, where the silane decomposition lifetime of about 0.1 sec is considerable longer than explosion times.

While it is almost never possible to deduce mechanisms based on stoichiometric results, the unique characteristics of silane-oxygen explosions provide an exception to this rule. We have proposed a mechanism[5] (shown below) that is consistent with the observations on these explosions and that is also consistent with our understanding of the thermochemistry of H,Si,O containing species.[5]

Mechanism of the Silane-Oxygen Explosion Reaction [a,b]

Low T

Reaction	
$R^{\cdot} + SiH_4$ --1-- $RH + SiH_3^{\cdot}$	$R = H^{\cdot}$ and $SiH_3O^{\cdot}$
$SiH_3^{\cdot} + O_2$ --2-- $(SiH_3O_2^{\cdot})^*$	$\Delta H^0 = -72$ Kcal
$(SiH_3O_2^{\cdot})^*$ --3-- $SiH_2OO + H^{\cdot}$	$\Delta H^0 = 62 \pm 10$ Kcal
SiH_2OO ---4--- $(HSiOOH)^*$	$\Delta H^0 = -108.2 \pm 10$ Kcal
$(HSiOOH)^*$ --5-- $^{\cdot}SiOOH + H^{\cdot}$	$\Delta H^0 = 90 \pm 2$ Kcal
$^{\cdot}SiOOH$ --6-- $SiO_2 + H^{\cdot}$	$\Delta H^0 = -77.1 \pm 2$ Kcal
$^{\cdot}SiOOH + O_2$ --7-- $SiO_2 + HO_2^{\cdot}$	$\Delta H^0 = 14.3$ Kcal
$H^{\cdot} + O_2$ ---8--- $HO_2^{\cdot}$	

High T

Reaction	
$SiH_4 + (M)$ --9-- $SiH_2 + H_2 + (M)$	$\Delta H^0 = 59 \pm 2$ Kcal
$:SiH_2 + O_2$--10-- $(H_2Si\binom{O}{O})^*$ -9'- $2H^{\cdot} + SiO_2$	$\Delta H^0 = -30.8$ Kcal

Reactions of the $H_2 + O_2$ Explosion

$H^{\cdot} + O_2$ --11-- $HO^{\cdot} + O^{\cdot}$
$O^{\cdot} + H_2$ --12-- $HO^{\cdot} + H^{\cdot}$
$HO. + H_2$ --13-- $H_2O + H^{\cdot}$

a) The mechanism above has been simplified from the original[5] by eliminating hot molecule reactions not producing H atoms.

b) H_2SiOO is the silicon analogue to the Criegie zwitterion prevalent in ozone + olefin reactions (i.e., H_2C=O-O).

Recent results support the main reactions of this mechanism. Thus Niki, et al.[8] have shown that the reaction of SiH_3 and O_2 produces H atoms according to the reaction, $SiH_3 + O_2$ --- $H^{\cdot}$ + H(OH)Si=O (which in our mechanism is the sum of reactions 2,3 and 5), and Horie, Potzinger and Reimann[9] report that H_2 was formed in a chain process when SiH_3 radicals were generated by Hg sensitized SiH_4 photolysis in the presence of small amounts of O_2.

One of the extraordinary characteristics of silane-O_2 explosions is that they can occur at temperatures as low as -80°C.[9] This strongly suggests the involvement of chemically activated 'hot' molecule reactions (e.g., rxn 3), as activation energies of only a few Kcal can be prohibitive to reactions of thermally equilibrated reactants at such low temperatures. By contrast, hot molecule spontaneous decompositions are mainly controlled by quantum state densities and sums (i.e., by entropy) and are not very temperature sensitive. RRKM calculations[5] of the spontaneous decomposition rate of $(SiH_3O_2^{\cdot})^*$ gives $k(3) \approx 10^{8.8}$ sec^{-1}. Collision rates ($k_{coll} = 10^6 \times P_{torr}$) are slower at pressures in the explosion regime (see below), hence the proposed hot molecule mechanism seems quite reasonable.

Our efforts to confirm reported upper pressure limits (P_2) for 30% SiH_4-70% O_2 mixtures by the Emeleus and Stewart method of pressure reduction[2] and by the Battelle method of temperature rise[3] were initially frustrated by data irreproducabilities. The early measurements of P_2 were significantly below the published values. However, it was eventually possible to demonstrate that the cause of this behavior was traces of water adsorbed on the Si and SiO_x reaction cell wall deposits. Scrupulous removal of all water from the reactants, reactant storage bulbs, reaction cell and connecting high vacuum lines resulted in P_2 values[5] (via both pressure reduction and temperature rise methods) in good agreement with the literature values[2,3]. By contrast, exposing a 'dry' reaction cell to 20 torr of water, then outgassing to a usual line pressure of 10^{-4} torr, followed by cell loading (with silane and O_2) and measurement of P_2 (by either the ΔT or ΔP method), always resulted in P_2 values well below those measured in the absence of water. Water, then, is a strong inhibitor of the silane-oxygen reaction, and adsorbed water on surface deposits is very hard to remove. Simple evacuation is not sufficient. The inhibitory effect of water may occur by way of H-atom scavenging as in reaction 14.

$$H^{\cdot} + H_2O \text{ =====14====== } H_3O^{\cdot}$$

Recent Results: Studies of the P_2 limits of 70% SiH_4-30% O_2 mixtures (not done before) under anhydrous conditions revealed that these mixtures exploded in ΔP studies on the first or second aliquating irrespective of the total pressure [suggesting shock-sensitivity], and exploded in temperature rise studies at 80 ± 10 C, again irrespective of the total pressure. Some data for these studies are shown in Table II. The pressure lowering results show that 70% SiH_4-30% O_2 mixtures are metastable, potentially explosive mixtures under all study pressure conditions and that they become explosive at T > 80 C. There must be some interesting mechanistic explanations for both the metastability of these silane rich-oxygen poor mixtures and for their easy transformations to the explosive condition. One plausible explanation, which follows from the proposed mechanism, involves silane trapping of the Criegie type intermediate, as in reaction 15, followed by the decomposition of its silylperoxide trapping product, as in rxn 16. Thus if SiH_2OO in the silane rich mixtures is scavenged by silane via rxn 15 before it can spontaneously decompose (and thereby continue and sustain a branching chain), the reactant mixture will not explode (although it is capable of doing so if the ratio of trapping to decomposition should vary to favor the latter). Thermal decomposition (rxn 16) of the silylperoxide at T > 80^0C (a reasonable possibility for such as compound) would then explain why explosions always occur at these temperatures. The silyloxy radicals produced in this decomposition are not only active radicals capable of chain propagation by abstraction from silane (rxn 17), but they are also capable of spontaneous rearrangement and decomposition to hydrogen, as shown in rxns 18,19.

$$^{\cdot}SiH_2OO^{\cdot} + SiH_4 \text{ ---15-- } SiH_3 00SiH_3$$
$$SiH_3OOSiH_3 \text{ ---16--- } 2\ SiH_3O^{\cdot}$$
$$SiH_3O^{\cdot} + SiH_4 \text{ ---17--- } SiH_3OH + SiH_3^{\cdot}$$
$$SiH_3O^{\cdot} \text{ ---18--- } (H_2SiOH)^{*} \text{ ----- } H + HSiOH$$
$$\text{----19---- } H_2 + SiOH$$

APPLICATION TO LABORATORY SILANE-OXYGEN EXPLOSIONS

The results in Table II appear to be the first experimental data that may relate to the erratic explosion problem mentioned earlier. Thus it is evident that metastable, but potentially very explosive, SiH_4-O_2 mixtures can be formed readily (as for example a slow silane leak into air under the 'right' kinds of conditions), and that very minor perturbations in these mixtures can produce explosions. [We have found, in our 70% SiH_4-O_2 mixtures, that a single light tap on the reaction cell was often sufficient to effect an explosion.] If there were a way to make silane-oxygen mixtures explode or flame more readily when their concentrations reached those of the explosion peninsula, the dangers of silane leaks would be significantly reduced (i.e. the highly concentrated conditions typical of the metastable mixtures would never be reached and the violent explosions typical of these mixtures would be replaced by either burning, which is controllable, or by milder and less damaging explosions typical of silane poor explosion mixtures. Since disilane (Si_2H_6) is more pyrophoric and explosive than silane, we felt that small additions of disilane to silane might act in the desired fashion (i.e., to suppress metastable mixture formation and to produce more spontaneously explosive mixtures). Also, disilane is an initial product of the silane decomposition, and such additions should have little or no effect on the industrial processes that rely on silane decomposition as a source of pure silicon. Thus we have studied the effects of additions of Si_2H_6 to SiH_4 as a function of percent Si_2H_6 in both static and dynamic flow experiments.

Static Studies: The experiments described in Table II were repeated with (70-X)% SiH_4, X% Si_2H_6, X= 4-8, in O_2 mixtures with results as shown in Table III. These studies show that additions of only a few percent disilane were sufficient to make the mixtures much more explosive.

Dynamic Studies: In these studies, silane or silane-disilane mixtures, in a 246 cc vessel at initial pressures around 200 torr, were leaked at controlled rates into a connecting vessel of 244 cc containing pure oxygen or air at initial pressures around 50 torr. The conditions of the resulting explosions (i.e., % SiH_4 just prior to explosion) were then determined from the pressure of the silicon hydride mixture prior to explosion. The latter was continuously monitored during the flow process with a pressure transducer. The objective was to produce conditions that were analogous to those of real-life silane leaks within the constraints of safety and control. The data from these experiments are shown in Table IV. Three results are clearly evident: 1) the addition of just a few percent disilane (2% and 4% for leaks into air and O_2, respectively) does significantly increase the probability of explosion, 2) the probability of explosion increases with increasing leak rate (irrespective of composition), and 3) the probability of explosion for leaks into air is higher than for leaks into oxygen. The latter observation demonstrates that O_2 can actually inhibit SiH_4-O_2 explosions under some conditions. This is consistent with rxn 8 of our mechanism and is presumably due to conversion of 'active' chain propagating H atoms into relatively inert HO_2 radicals.

The reason why small additions of disilane enhance the explosive character of silane-oxygen mixtures is not known, although because of the expected similarities in the chain reaction rates for silane and disilane oxidations, it is most reasonable to expect that the disilane rate enhancement is due to some change in the initiation reaction(s). For example, it is possible that silane-oxygen explosions are initiated homogeneously by the insertion of oxygen into SiH_4 to give silylhydroperoxide (rxn 20). Judging from the known behaviors of alkylperoxides and hydroperoxides, this species would be shock sensitive and produce in its decomposition (rxn 21) two 'active' radicals (SiH_3O and OH). Disilane is significantly more reactive toward silylene insertions into Si-H bonds than silane. Thus it is likely that disilane would be even more reactive to O_2 and produce similarly unstable explosion initiators (e.g., Si_2H_5OOH of SiH_3OOSiH_3 via rxns 22 and 23). Initiation in systems containing disilane, then, would be more facile.

$SiH_4 + O_2$ --20-- SiH_3OOH
SiH_3OOH --21-- $SiH_3O^{\cdot}$ + $^{\cdot}OH$
$Si_2H_6 + O_2$ --22-- SiH_3SiH_2OOH
$Si_2H_6 + O_2$ --23-- SiH_3OOSiH_3

Whatever the mechanism, it is clear that additions of small amounts of disilane to silane produces gas mixtures that are more explosive and pyrophoric in air than neat silane. It is therefore possible that such additions could solve the problem of silane leaks that cause major explosions. From our studies, it would appear that a large SiH_4-Si_2H_6 mixture leak into air should always ignite (or explode) readily, and this is desirable. By contrast, very slow leaks of SiH_4-Si_2H_6 mixtures should be safely handled by the rapid flow of air normally utilized in silane storage cabinets. The real problem arises at intermediate leak rates where explosions and burning occur only after significant buildups of silane have occurred. It is this kind of explosion that our studies indicate might be avoided by the addition of a few percent of disilane to the silane. However, since our results are dependent on the total pressure of the exploding mixtures (i.e., the higher the pressure, the less likely the explosion), to really assess the feasibility of this 'disilane solution' to violent explosions from silane leaks, studies similar to those reported here need to be done but with reactants at atmospheric pressures so as to obtain data under the normal CVD type flow conditions used in industry.

Acknowledgment: The authors are indebted to the Xerox Corporation for financial support of this work.

Table I: Stoichiometry of silane-O_2 explosion reactions [a]

% SiH_4	Overall Reaction/10 mol Reactant	-ΔH°(kcal)/10 mole
30	$3\ SiH_4 + 7\ O_2$ -- $3\ SiO_2 + 6\ H_2O + O_2$	600 [b], 1200 [c]
40	$4\ SiH_4 + 6\ O_2$ -- $3.2\ SiO_2 + 0.8\ SiO + 3.2\ H_2 + 4.8\ H_2O$	576 [b], 1096 [c]
50	$5\ SiH_4 + 5\ O_2$ -- $3.75\ SiO_2 + 1.25\ SiO + 8.75\ H_2 + 1.25\ H_2O$	429 [b], 1080 [c]
60	$6\ SiH_4 + 4\ O_2$ -- $1.7\ SiO_2 + 4.3\ SiO + 11.7\ H_2 + 0.3\ H_2O$	300 [b], 893 [c]
70	$7\ SiH_4 + 3\ O_2$ -- $6\ SiO + Si + 14\ H_2$	95 [b], 698 [c]
80	$8\ SiH_4 + 2\ O_2$ -- $4\ SiO + 4\ Si + 16\ H_2$	-268 [b], 493 [c]
90	$9\ SiH_4 + O_2$ -- $2\ SiO + 7\ Si + 18\ H_2$	-632 [b], 287 [c]

a) Initiation temperatures were 100^0C for 30 - 50% silane mixtures and 25^0C for 60-90% silane mixtures.
b) Reaction enthalpies calculated for all products as gases.
c) Reaction enthalpies calculated with SiO, SiO_2 and Si as solids and H_2O and H_2 as gases.

Table II Typical 70% SiH_4-30% O_2 mixture P_2 data; very dry conditions

Run	P_0[a]	P_2[b]	T(#C)[c]	*Method[d]	Run	P_0[a]	P_2[b]	T(#C)[c]	*Method[d]
1	150	167	39	ΔT	9	351	285	25	ΔP[e]
2	240	282	78	ΔT	10	240	170	25	ΔP[e]
3	248	282	79	ΔT	11	180	120	25	ΔP[e]
4	300	367	91	ΔT	12	50	30	25	ΔP[f]
5	300	351	75	ΔT	13	240	225	25	ΔP[f]
6	300	362	86	ΔT	14	351	285	25	ΔP[e]
7	180	212	78	ΔT	15	50	30	25	ΔP[f]
8	180	210	75	ΔT	16	150	110	25	ΔP[f]

a) Initial total pressure of reactant mixture (at 25 C).
b) Pressure of reactant mixture just prior to explosion.
c) Temperature of the reactant mixture just prior to explosion.
d) Methods of effecting explosion: ΔP (by pressure reduction); ΔT (by temperature rise).
e) Explosion occurred on the second opening to the expansion volume.
f) Explosion occurred on first opening to the expansion volume.

Table III P_2 Data for 70% silicon sydride-30% O_2 mixture explosions[a]

Composition %			Initial Pressure[a]	Explosion Condition[b]	
Si_2H_6	SiH_4	O_2	(in torr, 25 ^{0}C)	P_{torr}	T(^{0}C)
8	62	30	(151)	exploded in preparation	
6	64	30	(200)	exploded in preparation	
5	65	30	200	203	30
6	64	30	252	257	30
5	65	30	(150)	exploded in preparation	
4	66	30	150	153	31

a) Pressures in parenthese brackets are the pressures that would have been valid if the mixture had not exploded on warm up.
b) Explosions were effected by the temperature rise method.

Table IV Dynamic explosions, leak rate and disilane composition Effects

Initial Pressures Si_2H_6	SiH_4	O_2	Leak Rate (torr/min)	% Silane[a] at explosion	Explosion	#Runs
--	206	52	0.26 - 0.81	35 - 40	no	6
2	198	50	0.68 - 1.60	35	no	3
			1.80 - 2.00	23 - 30	yes	2
4	202	52	0.12	33	no	1
			0.41	32	yes	1
			0.30 - 0.67	15 - 19	yes	5
10	195	52	0.26	29	no	1
			0.52 - 0.73	8 - 13	yes	4
Si_2H_6	SiH_4	Air				
---	206	52	0.26 - 0.75	35	no	5
			1.10	11 - 16	yes	2
2	198	50	0.16	32	yes	1
			0.20 - 0.30	28	yes	2
			0.55	16	yes	1
			0.78	6	yes	1
4	202	52	0.07 - 0.13	11 - 20	yes	4
			0.16	7	yes	1
10	196	52	0.06 - 0.32	2 - 7	yes	7
			0.08	14	yes	1

a) % Silane in the mixture just prior to explosion
% Silane = silane added/(silane added + air or O_2 present)

REFERENCES

1) Friedel and Landenberg, Annalen 143, 124 (1867).
2) H.J. Emeleus and K. Stewart, J. Chem. Soc., 1182 (1935).
3) E.L. Merryman and A. Levy, Battelle Report (1980).
4) P.S. Shantarovich, Acta Physicochim., URSS, 6, 65 (1937).
5) J.R. Hartman, J. Famil-Ghiriha, M.A. Ring and H.E. O'Neal, Combustion and Flame, 68, 43 (1987).
6) J.H. Purnell and R. Walsh, Proc. Roy. Soc. A293, 543 (1966).
7) C.G. Newman, H.E. O'Neal, M.A. Ring, F. Leska, and N. Shipley, Int. J. Chem. Kinet., 11, 1167 (1979); J.W. Erwin, M.A. Ring, and H.E. O'Neal, Int. J. Chem. Kinet., 17, 1067 (1985).
8) H. Niki, P.D. Maker, C.M. Savage, and L.P. Breitenbach, J. Phys Chem., 89, 1752 (1985).
9) O. Horie, P. Potzinger and B. Reimann, to be published.
10) A. Stock and C. Somieski, Ber., B55, 396 (1922).

THE IMPORTANCE OF THE DIRECTION AND PROGRESS OF THE BUILDING AND FIRE CODES AND THEIR IMPACT ON THE PHOTOVOLTAIC INDUSTRY IN THE THREE REGIONAL AREAS OF THE UNITED STATES

Larry Fluer
Fluer, Inc., Carbondale, Colorado 81623

ABSTRACT

The codes that govern leading edge industries are in a state of change. Hazardous materials utilized by these industries have been the focus of concern for safety personnel of the industry and the general public. This concern has resulted in revisions to the codes affecting the users of hazardous materials whereby buildings once viewed as being acceptable are now viewed as being out of conformity to the codes, presenting both the building owner and the jurisdiction with a dilemma in that the building may no longer be legally operated or occupied. By using and understanding the codes developed for a sister industry, the semiconductor industry, the photovoltaic industry and the public will benefit.

INTRODUCTION

The building and fire codes across the country today include very specific provisions for the buildings and structures in which solar cells and other related devices are manufactured. Many of the hazardous materials and process techniques in use today presently parallel those employed by a sister industry-- the semiconductor industry. The codes written for "leading edge" industries provide welcome relief from what otherwise might result in severe restriction and costly retrofit for the type and size of buildings now occupied by the photovoltaics industry.

BACKGROUND TO CODE DEVELOPMENT

In 1980 regulatory authorities and the manufacturers of semiconductors and related devices realized that buildings built and occupied for 20 years or more as manufacturing and R & D occupancies were out of conformity to the codes. This non-conforming status occurred as the result of the lack of recognition that the industry was a user of relatively exotic, highly hazardous and highly toxic materials. These materials were present in quantities in excess of the exempt amounts then permitted by the codes for use in general manufacturing buildings, which were not considered to be highly hazardous. With the presence of hazardous materials recognized, primarily in the West, construction and retrofit were brought to a virtual standstill, as under the code, these facilities would have to be classified as a high hazard use; and continued operation and occupancy would be considered to be too a high risk, as well as a major liability for the jurisdiction and the owner should an accident or an incident occur.

The semiconductor industry reacted by undertaking a developmental effort in concert with the regulatory authorities to create a "code solution" which would provide safety for both the building occupants and the general public. A technical code solution rooted in scientific principles and applications was the focus of the work. The process involved the development of a "standard" which was based on a combination of construction features and engineering controls. This combination of controls was designed to produce a building environment that reduced the perceived risk, in semiconductor manufacturing and related research, to one no greater than that of a general manufacturing occupancy not using hazardous materials in excess of exempt amounts.

This code concept was revolutionary in its approach as it rejected the traditional approach to control based solely on confinement and compartmentalization in deference to quantity limitations, emergency shut-down, early detection, fire protection, material segregation, hazard isolation, alarms and notification.

A code concept was developed (1980-1984) in the Western Region, under the codes of the International Conference of Building Officials (ICBO) and the Western Fire Chiefs Association (WFCA). This concept was to become known as the "Group H, Division 6" occupancy.[1,2] This was followed, in succession, by the dominant model code in the Northeastern Region, the codes of the Building Officials and Code Administrators International (BOCA) with the " HPM (HPM - Hazardous Production Materials) Use Facility"[3]; and then in the Southern Region under the codes of the Southern Building Code Congress International (SBCCI) as the "HPM Facility"[4].

ICBO - 1985 H-6
BOCA - 1987 HPM Use Facility
SBCCI - 1988 HPM Facility

APPLICATION

The application of the code is determined by conformance to the following definition:

> SEMICONDUCTOR FABRICATION FACILITIES AND COMPARABLE RESEARCH AND DEVELOPMENT AREAS when the facilities in which hazardous production materials are used are designed and constructed in accordance with Section xxx and storage, handling and use of hazardous materials is in accordance with the Fire Code.

How may these codes be applied to existing buildings?

The occupancy or use classification of a building is determined based on the use to which a building or area of a building is placed. Manufacturing and R & D buildings have typically been classified as Group B, Division 2 occupancies in the West or as B or F use groups in the northern and southern areas of the United States. An exception to this exists for buildings which utilize hazardous materials which could either endanger building occupants or which exceed pre-established limits as stated in the codes.

If the materials utilized are hazardous, a question is then raised as to whether or not the occupancy should be classified as hazardous. The presence of the hazardous materials alone, has been justification for such a classification in the past, but not in the future, as the active state of the material will also have to be considered.

Buildings which have housed such materials, in excess of exempt amounts, have no protection under the so-called "grandfather provisions" of the code, as they likely fell out of conformity at some point when the exempt amounts were exceeded. This fact can have a severe impact on existing buildings by placing existing permits in jeopardy and by raising a liability issue for both the building owner and the jurisdiction.

If discovered, the jurisdiction has no real alternative other than to ask the building owner to remove those materials in excess or to revoke the owner's permits altogether.

What materials might the photovoltaics industry utilize that could create a code problem in existing buildings?

Typical of the materials categories most likely to be utilized by the industry where a problem is likely include:

Highly toxic materials and poisonous gas
Corrosive liquids
Flammable liquids and gases

Materials representative of these categories include:

Phosphine
Arsine
Diborane
Silane
Sodium hydroxide
Hydrogen
Sulfuric acid
Chromium trioxide
Arsenic
Isopropyl alcohol
Tetramethylstannane

Limitations on exempt amounts of these materials are determined by their classification and have been limited in the present codes to:

Highly toxic materials and poisonous gas - none
Corrosive liquids - 55 gallons
Flammable gases - 3,000 cubic feet
Flammable liquids (IB) - 120 gallons

The presence of these materials in excess of these exempt amounts means that the building or area which houses the materials must be classified as a highly hazardous building or area. Such classification would have the impact of reducing the maximum allowable building area for the typical building occupied to less than 10,000 square feet (total building).

The solution then becomes the creation of an H occupancy that allows existing buildings to function and grow while providing safety to the building occupants and to the general public.

Given that the "H-6" occupancy seems to be the appropriate occupancy, what requirements are likely to be imposed on users that may catch the user unaware?

Some of the more unusual requirements found in the H-6 occupancy include:

-The prohibition on the transport of hazardous materials in the exit corridor system.

-The need for emergency power on the exhaust ventilation system.

-Quantity limitations on the amounts of hazardous materials that may be utilized at work stations.

-The need for an upgraded sprinkler system (Ordinary Hazard 3).

-Mandatory gas detection systems with associated automatic shut-down of toxic gases.

-The prohibition on the mixing of general ventilation air with areas other than the manufacturing area which must be surrounded by a 1-hour occupancy separation.

-Limited travel distances to required exits and means of egress.

-Maximum density of HPM throughout the occupied area.

-The need for an on-site emergency response team with required drills and training records.

-The limitation on quantities of HPM that may be transported in either exits (existing buildings) or in service corridors (new construction).

-The need for excess flow protection and shut-off on HPM gases.

-The creation of a constantly attended "Emergency Control Station"

What trends are occurring within the model codes that will have an influence on the industry, and how may these trends affect the H-6 occupancy? Is there another occupancy that might be considered?

The development of the H-6 occupancy heightened the awareness on the part of the regulatory authority that current codes were lacking in their approach to control of hazardous materials.

Several years ago, the ICBO and the WFCA undertook a major effort to re-define the H occupancies, as well as to write an "improved" set of regulations for materials control by way of the fire code.

The trend toward more restrictive controls over hazardous materials with an increased acceptance of engineered systems as a means of control is apparent. The precedent set by the H-6 occupancy is having a far reaching impact on all occupancies handling hazardous materials.

Some of the recently developed concepts in the ICBO and WFCA codes include:

-The recognition of two major material hazard categories, those of Physical Hazard and those of Health Hazard.

-The creation of the H-7 occupancy to address those facilities using primarily health hazard materials, and which are relatively deplete of physical hazard materials.

-A relaxation on the prohibition of building areas and construction types for buildings built and maintained to the newly designed codes.

-Systems designed to address the episodic release of hazardous materials including chemical spill and gas leakage -- both inside and outside of buildings of concern. These systems include containment systems for the control of hazardous materials spills and of fire protection water in the event of such an incident as well as those required to contain and "treat" the episodic release of toxic gases.

-Creation of large buffered areas aroung buildings or processes containing "significant amounts" of material posing an explosion hazard (Silane, for example).

It may be predicted that in the West, the photovoltaics industry will occupy buildings housing mixed occupancies, with the occupancies being the H-6 and the H-7, depending on whether physical or health hazards are present, and in what quantity.

The model codes in the northeastern and southern United States (BOCA and SBCCI) are only now beginning to deal with the fact that, like the codes in the West (ICBO), these codes are deficient in the overall approach to hazardous materials control.

The BOCA "H-Use" committee is now in the initial stage of a major rewrite process of the Hazardous Occupancies section of the code. This rewrite will have a major impact on not only the photovoltaic industry, but all users of hazardous materials. Additionally, the SBCCI code committees are becoming aware of the issues and the problems related to the lack of control that the SBCCI codes exercise over the broad array of hazardous materials employed by industry in their geographic sphere of influence.

What will be the biggest impact by the codes on the industry, in the future?

Clearly, the pattern for controls has been set in the West by the ICBO. Other areas of the country are likely to follow and perhaps expand on this regional lead.

Existing buildings, and the controls exercised over the hazards contained therein, must be evaluated and understood by building owners now, in order to find areas of deficiency and to correct them before the fact. To fail to do so, or to delay, could cause the building to be declared non-conforming and operations either ordered to cease or to be changed radically. Should the realization of the problem occur after permits for expansion or remodel are being applied for, its discovery can have major financial repercussions and significant construction delays could be encountered.

What steps can be taken by those within the industry to ensure that the industry will grow and develop in a healthy code environment, resulting in buildings that are able to be occupied safely and at minimum risk to the building occupants and the public?

First, the industry must establish a philosophy toward the materials employed and the hazards perceived. Controls must be prudent, suited to the hazard, and proven in their ability to reduce risk and to minimize hazard.

Attempting to occupy a small lot or space to work with materials that require open expanses in event of mishap, for example, would be improper.

Failing to properly recognize the hazards or to thoroughly investigate the hazards of materials employed can not be tolerated. Generally, information from manufacturers is often incomplete and lacking. The industry has a responsibility to itself and to the public to be expert on the nature of all materials consumed, including physical and health properties as well as the short term and long term effects of personnel exposure or release into the environment.

Second, a corporate philosophy must be established that supports that of the industry. As corporate philosophies develop, and information is gained through experience, the acquired knowledge which produces a safer work environment must be shared for the benefit of the industry as a whole.

Third, the industry, as an industry, must participate in the on-going code development process. Facilities, safety, and environmental personnel must become active and stay informed of not only the current codes and regulatory climate, but of the trends and likely future impact of changes as well.

In any evolving technology, the needs of the process must be continually brought before those empowered to create regulation and impose controls. In today's regulatory environment, it is far too easy to become the victim of regulations conceived out of zeal, and not rooted in scientific and engineering principles, without the benefit of understanding the needs of the user or the end result of the regulation. Today, regulations impacting the use and construction of a building or its processes are created at the local civic level. These regulations are primarily based on the national and regional model codes and the standards of the industry. Failure of an industry to understand the codes and their impact, and to educate the authorities to the industry's needs can be costly and the results cannot be overstated.

SUMMARY

Because of heightened public awareness of hazardous materials, the codes that regulate the photovoltaics industry are in a state of flux.

Codes recently developed for the semiconductor manufacturers and comparable R & D activities have direct applicability to the photovoltaic industry due to the strong parallel between materials and processes used by that sister industry.

Many existing buildings permitted in the past may be found to be at odds with the code simply due to the restrictions placed on highly toxic materials, flammable gases, corrosives and other relatively exotic materials.

Code development in the Western Region is pointing the way to a solution that will provide greater building safety for building occupants and the general public.

The industry must develop a philosophy that addresses safety concerns related to building construction and control procedures for the facilities which it occupies.

The properties and the hazards of the materials employed must be thoroughly understood and related to appropriate code control measures.

Existing buildings must be constantly evaluated against newly developed codes to minimize risk and upset as the industry grows. Depending on yesterday's codes can lead to lack of appropriate and timely controls warranted by the materials contained and demanded by an awakening public.

REFERENCES

1. Uniform Building Code, International Conference of Building Officals, Whittier, California, 1985 Edition.

2. Uniform Fire Code, International Conference of Building Officals and Western Fire Chiefs Asssociation, Whittier, California, 1985 Edition.

3. The BOCA Basic Building Code/1987, Building Officals and Code Administrators International, Inc., Country Club Hills, Illinois.

4. Standard Building Code, Southern Building Code Congress International, Birmingham, Alabama, 1988 Edition.

THE APPLICABILITY AND EFFECT OF THE EMERGENCY PLANNING AND COMMUNITY RIGHT TO KNOW ACT ON THE PHOTOVOLTAICS INDUSTRY

Albert R. Axe, Jr. and Taryn-Marie McCain
Brown Maroney Rose Barber & Dye, Austin, Texas 78701

ABSTRACT

The purpose of this paper is to examine the applicability of the Emergency Planning and Community Right to Know Act of 1986 (EPCRA)[1] to the photovoltaics industry and briefly discuss a number of the major effects that EPCRA may have on that industry. Consequently, the paper begins with an overview of the four primary reporting requirements of EPCRA, emphasizing those aspects of these requirements that determine their applicability to a particular facility. The paper then discusses a number of the potential effects of EPCRA's applicability to the industry, concluding that well-reasoned and thorough planning is essential to responding to the law's new requirements.

INTRODUCTION

EPCRA was signed into law by President Reagan on October 17, 1986 as Title III of the Superfund Amendments and Reauthorization Act of 1986.[2] Drafted in part to help avert a Bhopal-type chemical release tragedy in the United States, EPCRA requires each state to form an emergency planning and response network, and requires facilities to provide information on the use of and accidental release of so-called "extremely hazardous substances" in order to facilitate prompt and effective emergency responses. EPCRA then goes further by establishing community right to know provisions, which require regulated facilities to disclose information on certain "hazardous chemicals" used, manufactured or processed, as well as routine releases of specified "toxic chemicals" into the environment. Detailed provisions relating to trade secret protection are also established by the Act.

While these facets of the legislation can be stated simply, both EPCRA and the regulations promulgated thereunder are very complex and often confusing. This brief article can thus only attempt to provide an overview of EPCRA and the issues it raises, and some general guidance for members of the photovoltaics industry who must comply with EPCRA.

APPLICABILITY OF AND NEW DUTIES CREATED BY EPCRA

A. EMERGENCY PLANNING FRAMEWORK AND REPORTING REQUIREMENTS

Sections 301-303 of EPCRA establish the general framework for emergency planning, define which facilities are subject to the emergency planning provisions, and create certain new reporting and information-sharing responsibilities for these facilities. Under § 301(a), each state was required to appoint a State Emergency

Response Commission (SERC) by April 17, 1987. The SERC was then to designate (by July 17, 1987) emergency planning districts within the state, and appoint (by August 17, 1987) for each district a Local Emergency Planning Committee (LEPC).[3] The LEPCs, which are to be supervised and coordinated by the SERC, were required to be composed of, at a minimum, elected officials, emergency response personnel, media, and representatives from community groups and affected industry.[4]

The most immediate role of these agencies is to develop and implement comprehensive plans for response to emergencies involving "extremely hazardous substances." Each LEPC is required to prepare such a plan by October 17, 1988, and submit it to the SERC for review to ensure coordination of LEPC plans statewide. The other major role of these agencies is to maintain, for review by the general public, copies of the emergency response plans and of the chemical information submitted to the agencies by regulated facilities.

EPCRA requires any "facility"[5] at which an "extremely hazardous substance" is present in excess of its "threshold planning quantity" (TPQ) to comply with its emergency planning requirements.[6] The list of extremely hazardous substances and their corresponding TPQs has been codified at Appendices A and B of 40 CFR Part 355.[7] This list includes many gases commonly used as deposition gases in the photovoltaics industry, such as phosphine, arsine, diborane, hydrogen selenide, hydrogen chloride, hydrogen sulfide, tetraethyltin, and trichlorosilane.

For TPQ purposes, the amount of an "extremely hazardous substance" present at a facility is the total amount of the substance present at any one time in concentrations of greater than one percent (1%) by weight, regardless of location, number of containers, or method of storage.[8] Special rules have been adopted by EPA for calculating the amount of an extremely hazardous substance present at a facility as a part of a mixture, solution or formulation.[9]

If a facility is subject to the emergency planning provisions of EPCRA, the owner or operator of the facility is required to do several things. First, the owner or operator of the facility was required to register with the SERC by May 17, 1987.[10] Thereafter, the owner or operator of a facility must register with both the SERC and the LEPC within 60 days after the facility first becomes subject to the emergency planning reporting requirements.[10]

Second, the owner or operator of a regulated facility must designate a facility representative to participate in the LEPC planning process as the facility's emergency response coordinator. The owner was required to have notified the LEPC of the appointment by September 17, 1987, or within 30 days after the LEPC is formed, whichever is earlier.[11]

Third, the owner or operator of a facility must, upon request by the LEPC, promptly provide to the LEPC "information ... necessary for developing and implementing the emergency plan."[12] The owner or operator must also promptly inform the LEPC of any changes occurring at the facility (including physical, operational, or management changes) which may be "relevant" to emergency planning, as such changes occur or are expected to occur.[13] Since the information which may be required is broadly defined, and the information which may be withheld as trade secrets is narrowly defined by EPCRA § 322, these provisions pose a significant potential for detailed inquiry by the LEPC into the operations at a facility.

B. HAZARDOUS CHEMICALS REPORTING

Sections 311 and 312 of EPCRA require certain reporting by the owner or operator of any facility required to prepare or have available a material safety data sheet (MSDS) for a hazardous chemical under the Occupational Safety and Health Act of 1970 and its implementing regulations.[14] Currently, the MSDS requirements adopted by the Occupational Safety and Health Administration (OSHA) apply to chemical manufacturers and importers (requiring them to assess the hazards of chemicals they produce or import), and manufacturing employers in SIC Codes 20 through 39 (requiring them to provide employees with information about the hazardous chemicals to which they may be exposed under normal conditions of use or in a foreseeable emergency).[15]

Under OSHA regulations, a "hazardous chemical" requiring an MSDS is a chemical which is a physical hazard (e.g., pyrophoric or reactive) or a health hazard (e.g., carcinogenic or toxic).[16] OSHA has not adopted a definitive list of hazardous chemicals, but deems hazardous those chemicals listed by OSHA and several health statistics organizations.[17] Unlisted chemicals must be evaluated by the chemical manufacturer or importer to determine whether they are hazardous.[18]

Under § 311, each facility subject to OSHA's MSDS requirements[19] must submit an MSDS for each hazardous chemical[20, 21] present at the facility (in at least the threshold amount for reporting set by EPA) to the SERC, the LEPC and the local fire department with jurisdiction over the facility. Alternatively, the facility may submit a list of the hazardous chemicals (grouped according to five "hazard categories" defined in 40 C.F.R. § 370.2), providing the chemical name or the common name for each chemical and any hazardous components of each chemical (as shown on the MSDS). However, if the facility submits a chemical list, the corresponding MSDSs must be submitted to the LEPC on request.[22]

Under § 312, each such facility must also submit an annual inventory of the amounts and locations of hazardous chemicals present at the facility in at least the minimum threshold amount set by EPA. The inventory must be submitted to the SERC, the LEPC and the local fire department with jurisdiction over the facility. Either a Tier I report (grouping chemicals into the five "hazard categories"), or a Tier II report (separate information per chemical) must be submitted containing:

(1) an estimate of the maximum amount of the hazardous chemical(s) present at the facility at any one time;

(2) an estimate of the average daily amount of the hazardous chemical(s) present at the facility (averaged using the number of days actually present[23]); and

(3) the location (Tier I - general, Tier II - specific) of the hazardous chemical(s) at the facility.[24] EPA adopted forms for Tier I and Tier II reports on October 15, 1987.[25]

While the owner or operator of a facility may choose to submit a Tier I rather than a Tier II report initially, he must submit a Tier II report upon request by the SERC, LEPC, or local fire department.[26] Further, upon request, the owner or operator of the facility must permit the local fire department to conduct an on-site inspection of the facility and must provide more specific information on the location of hazardous chemicals.[27]

EPA has adopted "minimum threshold levels for reporting" which will phase-in MSDS/list and annual inventory requirements for all hazardous chemicals except extremely hazardous substances.[28] The reporting threshold for extremely hazardous substances is set at a uniform 500 lbs. or the substance's TPQ, whichever is less, for the first and subsequent years of reporting.[29] The minimum reporting threshold for other hazardous chemicals is 10,000 lbs. for the first two reporting years, then drops to zero lbs.[30]

Under EPA's phasing schedule, MSDS (or list) submittals first became due October 17, 1987, for hazardous chemicals and extremely hazardous substances present in at least their threshold quantities as of that date,[31] and will be due for all remaining hazardous chemicals 2 years later.[32] Annual inventories will be due every March 1st, beginning in 1988, from facilities at which hazardous chemicals and extremely hazardous substances were present in at least their threshold quantities during the previous calendar year.[33]

Throughout phase-in and thereafter, there is a continuing obligation to supplement previous MSDS (or list) submittals. If an MSDS (instead of list) has been submitted, a revised MSDS must be submitted within 3 months of discovering significant new information regarding the chemical.[34] Also, an MSDS (or revised list) must be submitted within 3 months after the owner or operator is required to report on a new hazardous chemical.[35] The phase-in schedule will also not affect the requirement to submit an MSDS or Tier II report to the LEPC upon request. The threshold quantity in such instances is deemed to be zero.[36]

C. TOXIC CHEMICAL RELEASE REPORTING

Section 313(a) of EPCRA requires owners and operators of manufacturing facilities in SIC Codes 20-39 that employ ten or more full-time employees, and which manufacture, process[37] or use a specified list of toxic chemicals in excess of designated threshold quantities, to report routine releases of such chemicals on an annual basis to EPA and a state official (to be designated by the Governor by July 1, 1988).

The toxic chemicals covered by EPCRA § 313 consist of a list of 329 chemicals known as the "Maryland - New Jersey List" (a compilation of chemicals used for emissions reporting in those states).[38] The list contains a significant number of chemicals used in the photovoltaics industry, e.g., solvents (acetone, toluene, trichloroethylene, xylene, methanol) and thin film elements (copper, cadmium, selenium).

EPA published a uniform toxic chemical release form on June 4, 1987.[39] The information required, which is to be certified by a senior management official for accuracy and completeness, indicates for each toxic chemical (a) whether the chemical is manufactured, processed or used, and how it is used, (b) the estimated maximum amount present at the facility in the preceding calendar year, (c) the waste disposal/treatment method used and treatment efficiency (for each wastestream), and (d) the annual quantity entering each environmental medium.[40] The owner or operator may base the required information on readily available data collected pursuant to other laws (or if unavailable, reasonable estimates may be used), and is not required to conduct any special monitoring or measuring activities, or install any special devices.[41]

Only toxic chemicals manufactured, processed or used in excess of the threshold quantities set in EPCRA § 313(f) in a particular calendar year need to be included on the toxic chemical release report for that year.[42] Reports will be due on July 1st of each year beginning in 1988.[43]

D. EMERGENCY RELEASE REPORTING

The final major reporting requirement of EPCRA is contained in § 304 which requires the owner or operator of a facility[44] at which a "hazardous chemical" is produced, used or stored to notify the SERC and LEPC of certain releases of (1) EPCRA "extremely hazardous substances" and/or (2) "hazardous substances" subject to release reporting under § 103(a) of the Comprehensive Environmental Response, Compensation, and Liability Act of 1980[45] ("CERCLA").

To determine the applicability of this notification provision to a particular facility, the threshold question that must be resolved is whether the facility is one at which a hazardous chemical is produced, used or stored. The terms "facility" and "hazardous chemical" have previously been discussed in relation to other provisions of EPCRA.[46] A typical facility involved in the manufacture of photovoltaic devices would appear to be covered by § 304.

If the facility is subject to § 304, the next issue to be considered is whether a "release" has occurred. EPCRA § 329 broadly defines "release" to include virtually any entry of a defined substance into the environment, including the air, land, surface water or ground water. It is important to note, however, that EPCRA and the EPA rules which implement § 304 have created a number of important exemptions to the definition of release, including releases coming within the parameters of a federal permit and releases resulting in exposure to persons solely within the site(s) on which a facility is located.[47]

If a release from a covered facility has occurred, § 304 requires notification to be provided if a "reportable quantity" of an EPCRA "extremely hazardous substance" or a CERCLA "hazardous substance" has been released. "Extremely hazardous substances" have previously been discussed with respect to the emergency planning provisions of EPCRA §§ 301-303. The CERCLA "hazardous substances" are a separate list of substances (although overlaps exist between the two lists) codified in Table 302.4 of 40 C.F.R. Part 302. Both the list of hazardous substances (in Table 302.4) and the list of extremely hazardous substances (in Appendices A and B of 40 C.F.R. Part 355) indicate the "reportable quantities" of the various substances, i.e., that amount of the substance that must be released during any 24-hour period in order for a notification to be required.

Section 304 adopts the following general rules with respect to the notice that is required for a particular release:[48]

(1) If the release is reportable under CERCLA § 103(a), then the release must be reported to the National Response Center (under CERCLA), and the affected SERC(s) and LEPC(s), regardless of whether the substance is an EPCRA extremely hazardous substance.

(2) If the substance released is an extremely hazardous substance and is released in a EPCRA reportable quantity, but the release is not reportable under CERCLA § 103(a) [either exempt, not a CERCLA hazardous substance or not in CERCLA reportable quantity], then only the affected SERC(s) and LEPC(s) must be notified.

Notice to the National Response Center is given by telephone.[49] Generally, notice to a SERC or LEPC is to be given immediately by telephone, radio or in person, giving details of the release, its hazards, and proper precautions.[50] Written follow-up notice is to be submitted to the SERC and LEPC updating prior oral information and indicating response actions taken, health risks caused, and any advice on necessary medical attention.[51]

EFFECTS OF EPCRA ON PHOTOVOLTAICS INDUSTRY

From the above discussion of the new duties imposed upon the photovoltaics industry by EPCRA, it is obvious that the owners or operators of facilities in this industry are going to be required to comply with a plethora of new paperwork and reporting requirements. Compliance with these reporting requirements, which is enforceable, by substantial civil and criminal penalties and by citizens suits[52], will potentially have a profound effect on public knowledge of the chemicals used in the photovoltaics and other "high-tech" industries.

While the public may have previously known little about the types and characteristics of chemicals used by the photovoltaics industry, EPCRA emphasizes community awareness of this information. Under § 324 of EPCRA, each emergency response plan, MSDS (or list), annual inventory (Tier I or II), § 304 written follow-up and toxic chemical release form must be made available to the general public by the agency (EPA, SERC and/or LEPC) receiving the submittal.[53] Each LEPC is also required to publish annually in the newspaper the location at which these documents are available for review.[54] Only the location of specific chemicals reported in a Tier II annual inventory will be withheld from public disclosure upon the request of an owner or operator of a facility.[53]

EPCRA's disclosure requirements also threaten industry's ability to protect trade secrets. Under EPCRA § 322, an owner or operator may withhold only the specific chemical identity of any chemical required to be reported under § 303(d)(2) & (d)(3) (emergency planning information reported to the LEPC), § 311 (MSDS or list), § 312 (annual inventory), or § 313 (toxic chemical release report), as a trade secret.[55] To qualify as a trade secret, the chemical identity information must meet statutory requirements that (1) confidentiality has and will be protected, (2) no other laws require disclosure, (3) substantial competitive injury is likely if the information is disclosed, and (4) the identity is not readily discoverable through reverse engineering.[56] This protection of only chemical identity significantly narrows the protection traditionally afforded for trade secrets.[57]

Another more indirect impact of EPCRA on the photovoltaics industry may be in the area of toxic tort litigation. Perhaps the greatest barrier to a toxic tort plaintiff's ability to recover for harm allegedly caused by the release of chemicals from a facility is the issue of causation. In such suits, the causation factor can be broken down into two elements: "legal causation" (involving proof that it was <u>this</u> defendant's release of chemicals to which the plaintiff was exposed and not some other party's chemicals) and "medical causation" (involving proof that the exposure to these chemicals caused the illnesses or other damages of which the plaintiff complains). Much of the information needed to establish legal causation, including information on the chemicals present at a facility and emergency and routine releases of those chemicals to the environment, will now be readily accessible at the local level without the need to enter into lengthy and costly legal discovery procedures. In addition, the chemical release information required to be reported under §§ 304 and 313 of EPCRA should assist the toxic tort plaintiff in performing any necessary epidemiological studies used to establish proof of medical causation. Moreover, EPCRA § 313 and other provisions of the Superfund Amendments and Reauthorization Act of 1986[58] should serve as a catalyst to the production of more scientific data on the health effects of various chemicals, and thus assist future toxic tort plaintiffs to prove medical causation.

CONCLUSION

Due to the potential adverse effects of EPCRA on the photovoltaics industry, it is imperative that the industry responds carefully and thoughtfully to the new requirements of the Act. This includes carefully reviewing all EPCRA submittals with appropriate technical, legal and public relations staff, stressing, in particular, the protection of trade secrets. It also means developing a pro-active public relations program to educate the media and the public (especially public officials) about the industry and to respond effectively to media inquiries. Finally, due to the expanded notification and reporting requirements of EPCRA, it is extremely important that the industry continues to stress appropriate environmental safeguards in order to avoid chemical releases which must be disclosed to the public.

REFERENCES

[1]42 U.S.C.A. § 11001 et seq. (Supp. 1987).

[2]Pub. L. 99-499, 100 Stat. 1613 (1986).

[3]EPCRA § 301(b) & (c), 42 U.S.C.A. § 11001(b) & (c).

[4]EPCRA § 301(c), 42 U.S.C.A. § 11001(c).

[5]EPCRA § 329 defines "facility" to include all buildings, equipment, structures and other stationary items located on a single site or adjacent sites owned or operated by the same person. 42 U.S.C.A. § 11049(4). See also, note 45 hereto.

[6]EPCRA § 302(b)(1), 42 U.S.C.A. § 11002(b)(1). Note also that a Governor or SERC may designate additional facilities to be subject to emergency planning requirements. EPCRA § 302(b)(2), 42 U.S.C.A. § 11002(b)(2).

[7]The original list adopted by the U.S. Environmental Protection Agency (EPA) on November 17, 1986 (51 Fed. Reg. 41570 et seq.) consisted of 402 substances which EPA had identified as extremely hazardous in November 1985 as part of the agency's voluntary Chemical Emergency Preparedness Program. EPA subsequently added four substances and adjusted the TPQs of a number of other substances (52 Fed. Reg. at 13397, April 22, 1987) and delisted four other substances (52 Fed. Reg. 48072 and 48073, December 17, 1987).

[8] 52 Fed. Reg. at 13396 (1987) (to be codified at 40 C.F.R. § 355.30(a)).

[9]52 Fed. Reg. at 13395 (1987) (to be codified at 40 C.F.R. §§ 355.20 and 355.30(e)). See also, discussion in 52 Fed. Reg. at 13385 and 13392.

[10]EPCRA § 302(c), 42 U.S.C.A. § 11002(c); 52 Fed. Reg. at 13396 (1987) (to be codified at 40 C.F.R. § 355.30(b)).

[11]EPCRA § 303(d), 42 U.S.C.A. § 11003(d); 52 Fed. Reg. at 13396 (1987) (to be codified at 40 C.F.R. § 355.30(c)).

[12]EPCRA § 303(d)(3), 42 U.S.C.A. § 11003(d)(3); See also, 52 Fed. Reg. at 13396 (1987) (to be codified at 40 C.F.R. § 355.30(d)(2)).

[13]EPCRA § 303(d), 42 U.S.C.A. § 11003(d); 52 Fed. Reg. at 13396 (1987) (to be codified at 40 C.F.R. § 355.30(d)(1)).

[14]29 U.S.C.A. § 651 et seq. (1985 & Supp. 1987); 29 C.F.R. § 1910.1200.

[15]29 C.F.R. §§ 1910.1200(b)(1) & (2). Solar cell manufacturers fall within SIC Code 36. On August 24, 1987, OSHA extended the MSDS requirements to nonmanufacturers by requiring all employers to provide (by May 23, 1988) MSDSs on the hazardous chemicals to which their employees are exposed. 52 Fed. Reg. at 31877.

[16]See, definitions of "hazardous chemical," "health hazard," and "physical hazard" in 52 Fed. Reg. at 31879 (1987) (to be codified at 29 C.F.R. § 1910.1200(c)).

[17]See, 52 Fed. Reg. at 31879 (1987) (to be codified at 29 C.F.R. § 1910.1200(d)).

[18]52 Fed. Reg. at 31879 (1987) (to be codified at 29 C.F.R. § 1910.1200(d)(1)).

[19]OSHA exempts certain materials from MSDS requirements, including hazardous wastes subject to the Resource Conservation and Recovery Act (42 U.S.C.A. § 6901 et seq.) and manufactured articles. 52 Fed. Reg. at 31878 (1987) (to be codified at 29 C.F.R. § 1910.1200(b)(6)).

[20]EPCRA § 311(e) excludes (for §§ 311 and 312 purposes) five groups of substances from the definition of hazardous chemical, including any substance present as a solid in any manufactured item (if exposure to the substance would not occur during normal use), and any substance to the extent used in a research laboratory under the direct supervision of a technically qualified individual. 42 U.S.C.A. § 11021(e). See, 52 Fed. Reg. at 38347 regarding what constitutes a "research laboratory."

[21]See, 52 Fed. Reg. 38366 (1987) for special rules regarding mixtures (to be codified at 40 C.F.R. § 370.28).

[22]EPCRA § 311(c), 42 U.S.C.A. § 11021(c).

[23]52 Fed. Reg. at 38346 (1987).

[24]EPCRA § 312(d), 42 U.S.C.A. § 11022(d).

[25]52 Fed. Reg. at 38367 (1987) (to be codified at 40 C.F.R. §§ 370.40 & 370.41).

[26]EPCRA § 312(c), 42 U.S.C.A. § 11022(e).

[27]EPCRA § 312(f), 42 U.S.C.A. § 11022(f).

[28]52 Fed. Reg. at 38365 (1987) (to be codified at 40 C.F.R. § 370.20(b)).

[29]See, EPA preamble, 52 Fed. Reg. at 38352 (1987).

[30]EPA anticipates that it will revise the final threshold to be greater than zero lbs. 52 Fed. Reg. at 38350.

[31]See, EPA preamble, 52 Fed. Reg. at 38346 (1987).

[32]The phase-in schedule is adjusted for facilities which first become subject to the reporting requirements after October 17, 1987. See, 52 Fed. Reg. at 38365 (1987) (to be codified at 40 C.F.R. §§ 370.20(b)(1)).

[33]The three-year phase-in for annual reporting begins in the first year a facility becomes subject to such reporting requirements. See, 52 Fed. Reg. at 38365 (1987) (to be codified at 40 C.F.R. § 370.20(b)(2)).

[34]52 Fed. Reg. at 38365 (1987) (to be codified at 40 C.F.R. §§ 370.21(c)(1)). See, 52 Fed. Reg. at 38353.

[35]52 Fed. Reg. at 38365 (1987) (to be codified at 40 C.F.R. § 370.21(c)(2)).

[36]52 Fed. Reg. at 38365 (1987) (to be codified at 40 C.F.R. § 370.20(b)(3)).

[37]"Manufacture" and "process" are defined in EPCRA § 313(b)(1)(C), 42 U.S.C.A. § 11023(b)(1)(C).

[38]This list was published by EPA on February 4, 1987. 52 Fed. Reg. 3479. EPA, on its own initiative or upon petition by a member of the public or a State Governor, may add chemicals to or delete chemicals from the list by rule. EPCRA §§ 313(d) & (e), 42 U.S.C.A. §§ 11023(d) & (e). EPA outlined its policy on such petitions on February 4, 1987. 52 Fed. Reg. 3479.

[39]52 Fed. Reg. 21152.

[40]EPCRA § 313(g)(1), 42 U.S.C.A. § 11023(g)(1).

[41]EPCRA §313(g)(2); 42 U.S.C.A. § 11023(g)(2).

[42]EPCRA § 313(a), 42 U.S.C.A. § 11023(a).

[43]EPCRA § 313(f), 42 U.S.C.A. § 11023(f).

[44]For purposes of EPCRA § 304, "facility" includes motor vehicles, rolling stock and aircraft. EPCRA § 329, 42 U.S.C.A. § 11049.

[45]CERCLA, 42 U.S.C.A. § 9601 et seq. (1983 & Supp. 1987).

[46]See note 5 for the definition of "facility," and page 3 of the text for discussion of "hazardous chemical."

[47]EPCRA § 304(a)(2) & (a)(4), 42 U.S.C.A. § 11004(a)(2) & (a)(4); 40 C.F.R. § 355.40(a)(2). See, U.S.C.A. § 9603(a), (e) & (f) for releases exempt from CERCLA § 103(a) notification requirements.

[48]EPCRA § 304(a), 42 U.S.C.A. § 11004(a). These requirements vary for certain reportable CERCLA § 103(a) releases not involving extremely hazardous substances. See, EPCRA § 304(a)(3), 42 U.S.C.A. § 11004(a)(3).

[49]See, 40 C.F.R. § 300.36(c) (1986).

[50]EPCRA § 304(b), 42 U.S.C.A. § 11004(b); 52 Fed. Reg. at 13396 (1987) (to be codified at 40 C.F.R. §§ 355.40(b)(1) & (2)); Different rules apply to notice of releases involving transportation. See, EPCRA § 304(d), 42 U.S.C.A. § 11004(d).

[51]EPCRA § 304(c), 42 U.S.C.A. § 11004(c); 52 Fed. Reg. at 13396 (1987) (to be codified at 40 C.F.R. § 355.40(b)(3)).

[52]See, EPCRA §§ 325 & 326, 42 U.S.C.A. §§ 11045 & 11046.

[53]EPCRA § 324(a), 42 U.S.C.A. § 11044(a).

[54]EPCRA § 324(b), 42 U.S.C.A. § 11044(b).

[55]EPA proposed procedures for claiming trade secret status, and EPA review of such claims, on October 15, 1987. 52 Fed. Reg. 38312.

[56]42 U.S.C.A. § 11042(b).

[57]EPA has interpreted this section as impliedly excluding claims of the trade secrecy or business confidentiality of other information collected under sections 303(d)(2) & (d)(3), 311, 312, and 313. See, Proposed Rule 40 C.F.R. § 350.3(c)(1), 52 Fed. Reg. at 38325 (1987). EPCRA does not expressly allow (and may thus disallow) withholding of chemical identity (as a trade secret) from Section 304 emergency release notifications.

[58]See, SARA § 110, 42 U.S.C.A. § 9604(i) (Supp. 1987).

A NEW ENFORCEMENT ERA: INDIVIDUAL LIABILITY UNDER ENVIRONMENTAL STATUTES

John W. Teets
Division Counsel
Motorola, Inc.

ABSTRACT

This paper examines the applicability of environmental criminal sanctions to individuals engaged in the handling of hazardous substances in the photovoltaics industry. The paper begins with an overview of recent governmental enforcement action which is followed by an analysis and discussion of recent court decisions imposing liability on corporate employees and officers. The paper then examines the enforcement mechanism of the major environmental statutes and concludes with an outline of a corporate program to minimize exposure to governmental enforcement actions.

INTRODUCTION

There are numerous safety, health and environmental concerns in the production of photovoltaic cells. As in the semiconductor industry, hazardous gases are used in large quantities as major feedstock materials in most thin film photovoltaic cell technologies.[1] The highly toxic nature of these gases and the large volumes required for manufacturing purposes make it imperative that professionals in the photovoltaic industry command a competent and complete understanding of local, state and federal environmental requirements. This is especially critical given recent governmental enforcement action against individuals and the continuing criminalization of environmental harms.

The National Survey of Crime Severity conducted by the Department of Justice ("DOJ") highlights the public visibility associated with environmental crimes. The poll was conducted over a six-month period by the Center for Studies in Criminology and Criminal Law, Wharton School, University of Pennsylvania. The 60,000 participants ranked a pollution incident resulting in deaths as the seventh most severe crime, ahead of crimes like skyjacking and drug smuggling. Pollution of a city's water system in which there were no specified adverse health results ranked ahead of roughly 70 other traditional crimes.

The public has become much more aware of environmental hazards as a result of publicity associated with the Love Canal, Three Mile Island, and Bhopal, India incidents. Congress and the Courts have responded to public pressure concerning environmental harms. Congress has placed more emphasis on criminal prosecutions as a deterrent to environmental offenses and has given the government additional statutory provisions for the imposition of criminal sanctions.[2]

Virtually all local, state and federal environmental laws, regulations and ordinances contain enforcement provisions. In many cases these provisions include specific civil and criminal provisions available against corporate employees. Large fines and lengthy imprisonment are often specified for so-called "knowing" violations. These sanctions have been imposed upon corporations and their employees who incorrectly handle and dispose of hazardous substances and wastes even though traditional criminal conduct is absent and the defendants may be unaware of legal requirements.

A. RECENT ENFORCEMENT TRENDS

The Reagan Administration's enforcement policy, although less vigorous overall than that of previous administrations, has concentrated resources on personal, criminal enforcement. Whatever the reasons for past reluctance to prosecute violators of environmental statutes, criminal enforcement has become a priority. Originally, industry was adjusting to a novel, complex and often confusing environmental regulatory scheme. The recent large fines and lengthy prison sentences are indicative of a governmental policy based on the belief that the environmental statutes have been around long enough to be familiar to all in the regulated community thus making "knowing" violations truly criminal in nature.

The government has aggressively imposed civil and criminal penalties for violations to erase any economic or competitive advantage and to deter personal conduct. The DOJ's statistical record illustrates this evolution of enforcement policy. Throughout the 1970s, only 25 criminal cases were prosecuted. In contrast, from 1981 through 1985 DOJ filed 852 civil enforcement complaints and 600 judicial consent decrees. Most notably, from 1983 to 1987 DOJ successfully obtained 339 criminal indictments. As of September 30, 1987, 262 convictions or guilty pleas have been entered.[3]

A June 25, 1987 internal memorandum indicates that since DOJ's Environmental Consensus Unit was established in 1983 there has been a significant increase in the number and success of environmental criminal prosecutions. The following table illustrates this activity:

YEAR	PLEAS/ INDICTMENTS	CONVICTIONS	FINES	JAIL TERMS
1983	40	40	$ 341,100	11 years
1984	43	32	384,290	5 years 3 months
1985	40	37	565,850	5 years 5 months
1986	179	150	1,917,602	124 years 2 months
1987	84	43	2,512,000	24 years 6 months
TOTAL	386	302	$5,1720,842	170 years 19 months

Indictments against individuals include corporate officers, directors and employees. Currently, more than 47 years of actual jail term time have been served.[4]

Because of this potential exposure to both personal and corporate punishment is so great, it is important that all persons responsible for hazardous materials handling and waste disposal appreciate the full range of enforcement options available to the government and the substantive factors that characterize an environmental criminal case.

B. STANDARDS OF LIABILITY

Generally, criminal acts require an element of intent, called mens rea ("guilty mind") or scienter ("criminal intent"). This is also true in the environmental criminal provisions highlighted in the following section. These statutes require an act to be done "knowingly" and/or "willfully". However, as discussed in the introduction to this paper, the legal definition of these terms is quite different from that used by a non-lawyer.[5] Furthermore, in criminal prosecutions under public welfare statutes the government faces a less stringent burden proving intent than in traditional criminal statutes. Environmental statutes are enforced much like the Food and Drug Act a traditional public welfare statute. In these statutes Congress endeavored to control hazards that in the circumstances of modern industrialism are largely beyond "self protection." Courts have determined that it would undermine the purpose of the legislation to limit liability to those who only had actual knowledge of the requirements when the handling of hazardous materials permeates our society. Therefore, recent case law interprets the "knowing" element as requiring knowledge of the action and not the law. This interpretation functionally imposes a strict liability standard.

A corporation may be held liable for the criminal acts of its officers, directors and employees acting on behalf of the corporation within the scope of their employment. Similarly, corporate officers, directors and employees may be held responsible for their own criminal acts performed within the scope of their employment. The DOJ and EPA position concerning criminal enforcement highlights this principle:

> It is self-evident that corporations work through individuals and that the nature of criminal responsibility is such that it properly falls on individuals. The Justice Department will attempt to identify the individuals responsible for the corporate acts so that the law may be truly enforced and its real deterrent effect mobilized.[6]

An individual, including a corporate employee may be held criminally liable even if he did not know that his acts were illegal because of the public welfare status of environmental statutes. The United States Supreme Court has gone even further and concluded that those who deal with hazardous substances are expected to inquire into and know the law:

> [Where] obnoxious materials are involved, the probability of regulation is so great that anyone who is aware that he is in possession of these or dealing with them must be presumed to be aware of the regulation.[7]

The situation in the case of United States v. Hayes International Corporation[8] demonstrates the principal of imputed knowledge for meeting the "knowing" requirement. Hayes refurbished airplanes and in the course of doing so, generated wastes, including fuel drained from fuel tanks and solvents used to clean paint guns. L. H. Beasley, a Hayes employee responsible for disposing of hazardous wastes, negotiated a contract with a recycler who agreed to "take care of" the spent solvents. The spent solvents were subsequently discovered at an illegal disposal site and the U.S. Attorney brought a criminal indictment against Mr. Beasley and others. Mr. Beasley argued that he thought all wastes sent to a recycler were exempt from regulation, even if the wastes were not actually recycled. The Courts' response was simply that ignorance of the law is no excuse. The court went on to say, however, that the deal was such a "good deal" that Mr. Beasley should have been suspicious and should have inquired into the actual disposal arrangements. Accordingly, the court held that Mr. Beasley's lack of knowledge of the law as of disposal details should not absolve him from criminal responsibility for the illegal disposal.

In a slightly different vein, middle and upper level managers may be held liable for certain criminal acts committed by employees handling hazardous materials under their supervision. For example, in United States v. Johnson & Towers, Inc.,[9] workers at an automobile repair shop dumped degreasers and other chemicals into a trench which emptied into the Delaware River. The U.S. Attorney issued an indictment naming the service manager and foreman as defendants. This case is significant from two standpoints. First, the court stated that the service manager and foreman, as managers, could be held liable for the certain criminal acts committed by employees under their supervision. Second, as middle managers, the defendants were not necessarily the employees within the company responsible for obtaining the necessary permits. The defendants in Johnson & Tower argued that they were not even in a position to know whether or not the company had the necessary permits. Nevertheless, the court concluded that a person who manages a portion of a company's hazardous waste disposal activities should be aware of the permitting status of the company. Accordingly, the manager's lack of knowledge would not absolve him from criminal liability.

As the above cases indicate, a person who deals with hazardous substances is charged with the responsibility of knowing the law and inquiring into suspicious situations. If a person neglects to inquire, his lack of knowledge will not absolve him from criminal liability. Thus, the "knowing" prerequisite for criminal provisions in environmental statutes is met merely by proving the defendant knew or was aware of his actions and not necessarily the applicable legal requirements.

A more complex situation arises where a responsible corporate officer avoids knowledge of his subordinate's actions. Although corporations can be held criminally liable based upon the imputed knowledge of their employees, and employees may be held personally liable for "knowing" violations even though they are not aware of regulatory requirements, the general rule is that there is no implied knowledge of the criminal act to individuals in criminal cases. In other words, the individual must be aware of the act itself. However, prosecutors have increasingly employed two doctrines to reach higher level corporate officials. The first is the "avoidance of the truth" doctrine where the defendant deliberately avoided learning the truth so that proof of actual knowledge or willfulness could not be made. The second, the inferential doctrine allows a jury to infer knowledge or willfullness based on various kinds of evidence including (1) the defendant's position in the corporation; (2) applicable corporate policies; and (3) the existence of the regulatory scheme. From these factors courts permit an inference that, by virtue of position in the corporation and reporting and compliance responsibility, the defendant knew what he would have known through the exercise of due diligence.[10] These doctrines may also be referred to as the responsible corporate officer theory. Despite the confusing and inconsistent titles, the principle is the same for responsible corporate officers under public welfare statutes where prosecutors need not prove that the official charged with a criminal violation had actual knowledge of the violation. A good example of the exercise of this principle is United States v. A.C. Lawrence Leather Co.,[11] where five corporate officials were indicted under several criminal charges, including Clean Water Act violations. The company regularly bypassed its wastewater treatment facility and dumped raw waste directly into a local river. Two of those indicted, the president and vice president of the company, were indicted because they failed to seek out and halt the illegal practices. Similarly, in Gordon v. United States,[12] the court allowed partners to be convicted of willful violations based on the knowledge of lower level employees.

D. ENFORCEMENT PROVISIONS OF VARIOUS ENVIRONMENTAL STATUTES

The Clean Air Act criminalizes "knowing" violations of specified provisions by "any person," which is defined to include any "responsible corporate officer."[13] The Act imposes penalties of up to $25,000 per day of violation and/or incarceration for up to 1 year for the first offense. Second or subsequent offenses are punishable by fines of up to $50,000 and/or 2 years incarceration. The Clean Air Act also criminalizes knowingly making a false statement, with a fine of up to $10,000 and/or up to 6 months imprisonment.

The Clean Water Act criminalizes "willful" or "negligent" violation of effluent limitations or permit conditions by "any person," defined to include "any responsible corporate officer."[14] The Act imposes fines between $2,500 and $25,000 per day and/or up to 1 year incarceration for the first offense. Fines of up to $50,000 per day and/or up to 2 years incarceration are imposed for second or subsequent offenses. Section 309 of the Act also criminalizes the making of false statements. Penalties of up to $10,000 and/or incarceration for up to 6 months are imposed for the making of false statements.

In United States v. Frezzo Brothers, Inc.,[15] the defendants included the corporation and individuals who owned or were officers of the company. The defendants were convicted of willfully or negligently discharging runoff from composting operations. Section 311(b)(5) of the Act punishes the failure to provide notice of discharge of oil or hazardous substances with a fine of up to $10,000 and/or up to 1 year imprisonment.

The Resource Conservation and Recovery Act ("RCRA"), § 3008(d) criminalizes a variety of knowing violations regarding transportation of waste to unpermitted facilities, treating, storing, or disposing of waste without a permit, making false statements, and transporting wastes without a permit.[16] Generally, RCRA violations are punishable with a fine of up to $50,000 per day and/or up to two years incarceration. Transportation of hazardous waste to an unpermitted facility or treatment, storage, or disposal without a permit is punishable with a fine of up to $50,000 and/or up to 5 years incarceration. For second or subsequent offenses, RCRA penalties are doubled.

RCRA § 3008(e) creates a separate crime entitled "knowing endangerment." Any person whose knowing violation of § 3008 places another person in imminent danger of death or serious injury may be fined up to $250,000 and/or imprisoned for up to 15 years. A corporation convicted for knowing endangerment may be fined up to $1 million. Again, the knowing requirement only requires that the actor be aware of his actions and not the fact that the actions are prohibited. Everett Harwell, a former president of Southeastern Waste Treatment was fined $20,000 and sentenced to 3 years in prison for disposing of hazardous waste without a permit and making false statements. Eugene Dagett, a former vice president of Southeastern, was sentenced to 1-1/2 years and $10,000 fine for storing and transporting hazardous waste without a permit.[17]

Section 103 of the Comprehensive Environmental Response Compensation and Liability Act ("CERCLA") makes any "person" (defined to include "individuals") liable for failure to provide notice of the release of a hazardous substance from a vessel or facility.[18] CERCLA imposes penalties of up to $10,000 and/or up to 1 year incarceration for such a failure. Section 103 also criminalizes the knowing destruction or falsification of records, with a fine of up to $20,000 and/or up to 1 year imprisonment.

Title III of the recently enacted Superfund Amendment Reauthorization contains a freestanding provision entitled the Emergency Planning and Community Right-to-Know Act.[19] Section 325 of the Act penalizes knowing and willful failure to comply with the emergency release reporting provisions of § 304. Title III imposes fines up to $25,000 and/or up to 2 years incarceration for first offense and fines of up to $50,000 and/or 5 years imprisonment for second or subsequent offenses.

The Toxic Substance Control Act criminalizes knowing or willful violation of the prohibitions listed in § 15 of the Act.[20] Penalties include fines of up to $25,000 and/or incarceration for up to 1 year.[21] O. Lynwood Holley was sentenced and his corporation fined $60,000 for violating PCB disposal regulations.[22] Three officials of Drum Recovery Inc. were sentenced for illegal transportation of PCBs and sodium hydroxide. Sentences included entering an alcohol abuse program, fines, probation, community service and occupational disqualification.[23]

There are numerous non-environmental criminal laws available for environmental criminal prosecution that pose an even more credible threat when considered against the backdrop of the environmental laws' massive information requirements.

The Federal Criminal Code makes knowingly and willfully false statements to federal government subject to fines up to $10,000 and/or imprisonment up to 5 years.[24] The Criminal Code penalizes use of mails, airwaves, or interstate wires in connection with a "scheme or artifice" to defraud or for obtaining money or property by means of false or fraudulent representations.[25] In United States v. Gold,[26] a chemical manufacturer and its officers were indicted for making false statements to EPA. The crime of conspiracy will be available to enforcement officials if two or more corporate employees conspire to violate environmental laws.[27] Penalties include fines up to $10,000 and/or imprisonment up to 5 years. In United States v. Olin Corp.,[28] defendants were charged with a violation of conspiracy law and the false statements provisions of the criminal code in connection with plans to defraud EPA regarding mercury discharges into the Niagara River.

CONCLUSION

Typically, the government will not seek criminal enforcement unless the case either involves: (1) operating or doing business without a permit, or with a facility lacking a permit; or (2) the making of false statements. Both of these factors involve action which undermines and erodes the regulatory program by causing competitors to perceive an unfair competitive advantage through noncompliance with costly environmental requirements.

The first factor must be put in the context of the Hayes case to appreciate its breadth, especially when the individual is involved in the handling of hazardous substances and wastes. Again, the culpability requirement of "knowing" is a legal pitfall which will entrap the unwary in the environmental context because knowledge can be imputed based upon corporate position or the employee's responsibility for hazardous substances.[29]

The second factor -- making knowingly false statements -- strikes at the heart of environmental regulatory programs by undermining their integrity. "EPA's nationwide regulatory program can only function effectively with widespread voluntary compliance and honest self-reporting."[30] For this reason, concealment and false reporting only compound a violation and will be a prime target for criminal enforcement.

Situations lending to governmental enforcement actions can be avoided by an institutionalized management structure and corporate policy committed to environmental compliance.[31] This formula contains the following components:

1. Develop compliance plans and programs which communicate upstream to senior management and downstream to employees via training programs and refresher courses;

2. Adequately implement these plans with care given to reporting levels within the organization and employee experience;

3. Document a suitable degree of diligence and care concerning these programs by responsible high level officials;

4. Delineate lines of responsibility;

5. Develop and use compliance and procedures manuals;

6. Identify potential problems through independent or internal audits which permit the company to address the problems on its own terms; and

7. Adopt a written corporate policy regarding environmental offenses and the right to defense costs.

REFERENCES

1. V.M. Fthenakis, P.D. Moskowitz and L.D. Hamilton, Personal Safety in Thin Film Photovoltaic Cell Industries, Paper presented at the SERI Photovoltaics Safety Conference, January 16-17, 1986, Lakewood, CO.

2. See infra Note 29.

3. F. Henry Habicht II, The Federal Perspective on Environmental Criminal Enforcement: How to Remain on the Civil Side, 17 Environmental Law Reporter 10478 (Dec. 1987).

4. The source of statistics is a June 25, 1987 internal DOJ Environmental Crimes Section Memorandum.

5. J.G. Arbuckle and Paul A.J. Wilson, Observations on Developing Philosophy of Individual Criminal Liability for Environmental Law Violations, Paper presented at the Governmental Institute Environmental Laws & Regulations 1987 Update Conference, October 22-23, 1986, Washington, D.C.

6. Robert M. Perry, Unpublished EPA internal memorandum, Criminal Enforcement Priorities for the Environmental Protection Agency (Oct. 12, 1982).

7. _United States v. International Minerals & Chemical Corporation_, 402 U.S. 558, 565 (1971) (emphasis added).

8. 786 F.2d 1499 (11th Cir. 1986).

9. 741 F.2d 662 (3d Cir. 1984).

10. Supra Note 2.

11. No. Cr. 82-00037 (D.N.H. Sentencing Apr. 29, 1983).

12. 347 U.S. 902, 910 (1954).

13. 42 U.S.C. §§ 7401-7642.

14. 33 U.S.C. §§ 1251-1376.

15. 461 F. Supp. 266 (E.D. Pa. 1978) _aff'd_ 602 F.2d 1123 (3d Cir. 1979) _cert denied_ 404 U.S. 1074 (1980).

16. 42 U.S.C. §§ 6901-6987.

17. 1 Toxics Law Reporter 897 (Jan. 21, 1987).

18. 42 U.S.C. §§ 9601-9675.

REFERENCES - Continued

19. Pub. L. No. 99-499, 99th Cong., 2d Sess. (1986).

20. 15 U.S.C. §§ 2601-2616.

21. See Generally, Elizabeth Ann Hurst, Toxics and Enforcement, paper presented at the State Bar of Texas Institute Conference on Enforcement of Environmental Laws, December 5-6, 1985, Dallas, Texas.

22. 14 Env't. Rep. (BNA) 1483 (Dec. 23, 1983).

23. 15 Env't. Rep. (BNA) 179 (June 8, 1984).

24. 18 U.S.C. § 1001.

25. Mail Fraud 18 U.S.C. § 1341 and Wire Fraud 18 U.S.C. § 1343.

26. 470 F. Supp. 1336 (N.D. Ill. 1979).

27. 18 U.S.C. § 371.

28. 465 F. Supp. 1120 (W.D.N.Y. 1979).

29. Debra Mitchell and John Teets, Environmental Liability, 5 Motorola Environmental Newsletter (2nd Quarter, 1987).

30. Supra Note 3 at page 10482.

31. See Generally note 5 Supra.

CODES AND REGULATIONS GOVERNING HANDLING, STORAGE, USE AND RELEASE OF TOXIC GASES

Richard A. Bolmen, Jr.
Siliconix inc., Santa Clara, CA

ABSTRACT

To control the hazards associated with the handling, storage, use and release of toxic gases in the manufacturing process, a number of codes and regulations have been promulgated. At the forefront of these regulations are the Uniform Fire Code (UFC), Article 51, "Semiconductor Fabrication Facilities Using Hazardous Production Materials", 1985 Edition; UFC Article 80, "Hazardous Materials", 1987 Revision; and the Toxic Gas Model Ordinance, California Assembly Bill 1021. Article 51 of the UFC is specific for semiconductor wafer fabrication facilities and regulates the storage, handling and use of hazardous production materials. Article 80 of the UFC has a much broader scope and incorporates requirements for prevention, control and mitigation of dangerous conditions related to hazardous materials. As it applies to toxic gases, Article 80 incorporates Article 51 requirements and in addition requires treatment systems to reduce discharge concentrations and mitigate unauthorized releases of toxic gases. The Toxic Gas Model Ordinance is specific to toxic gases and regulates storage, handling and use of toxic gases at new and existing facilities incorporating monitoring and treatment systems should an unauthorized release occur. This paper will focus on these regulations as they apply to toxic gases used by the semiconductor and photovoltaics industries.

INTRODUCTION

The last five years have seen dramatic changes in the regulations governing toxic gas storage, dispensing, monitoring and control. Concerns over catastrophic leaks triggered by toxic gas releases at Bhopal, India and Institute, West Virginia have focused the regulatory and public sectors' attention on industries using toxic gases in their manufacturing processes. These concerns have been addressed in a number of forums and have resulted in many new codes and regulations aimed at proper engineering and administrative controls to prevent unauthorized gas releases, assess worst case scenarios via dispersion modeling, mitigate gas releases to a safe level via scrubbing/abatement systems and provide a higher level of community awareness via community "Right-to-Know" programs and notification systems (Table I).

Table I Scope of Existing & Emerging Regulations

UBC Section 9.911 Division 6 Occupancies (1985): Defines building code requirements for wafer fabs using hazardous production materials (HPM's) in excess of Table No. 9-A of the Uniform Building Code.[1]

UFC Article 51 (1985): Regulates storage, handling and use of HPM's in buildings in which HPM's are used in excess of Table No. 9-A, UBC.[2]

UFC Article 80 (1987 revision): To provide requirements for the prevention, control and mitigation of dangerous conditions related to specific classes of hazardous materials and define information needed by emergency response personnel.[3]

Toxic Gas Model Ordinance (1987): Protection of health, life, resources and property through prevention and control of unauthorized releases of toxic gases.[4]

CODE DEFINITIONS OF TOXIC GASES

The main focus of existing codes and regulations (UFC Article 51 & UBC 9.911/H-6) has been hazardous production materials (HPM's). An HPM is a solid, liquid or gas that has a degree-of-hazard rating in health, flammability or reactivity of Class 3 or 4 as ranked by Uniform Fire Code (UFC) Standard No. 79-3, and which is used directly in research, laboratory or production processes which have as their end products materials which are not hazardous.[5] Toxic gases fall within this classification (Table II). In addition to Uniform Building Code (UBC) and Uniform Fire Code (UFC) requirements, hazardous gases used by semiconductor manufacturers are regulated by insurance underwriting requirements, local fire codes and municipal ordinances. These are beyond the scope of this article, but should be incorporated into any hazardous materials/toxic gases management plan. Beyond the inclusion of highly toxic materials and poisonous gases as a class of HPM, existing codes do not provide a clear definition of a "toxic gas". However, many of the engineering controls (excess flow valves) and monitoring requirements clearly target (but, are not exclusive of) the highly toxic metal anhydride gases. These gases are used as dopants in concentrations from low ppm to 100% and represent the greatest hazard potential to employees and the surrounding community.

Table II Hazards and Exposure Limits for Toxic Gases

Gas	Chemical formula	Hazard Class*	TLV-TWA[6] (ppm)
Arsenic pentafluoride	AsF_5	T	not listed
Arsine	AsH_3	T,F	0.05
Boron trichloride	BCl_3	T,C	1
Boron trifluoride	BF_3	T,C	1
Carbon monoxide	CO	T,F	50
Chlorine	Cl_2	T,C	1
Diborane	B_2H_6	T,F,P	0.1
Ethylene oxide	C_2H_4O	T,F	1
Fluorine	F_2	T,C,F	1
Germane	GeH_4	T,F	0.2
Hydrogen bromide	HBr	T	3
Hydrogen fluoride	HF	T,C	3
Hydrogen selenide	H_2Se	T,F	0.05
Hydrogen sulfide	H_2S	T,F	10
Nitric oxide	NO	T	25
Nitrogen trifluoride	NF_3	T	10
Phosgene	$COCl_2$	T	0.1
Phosphine	PH_3	T,F,P	0.3
Phosphorus pentafluoride	PF_5	T,C	not listed
Phosphorus trichloride	PCl_3	T,C	0.2
Phosphorus trifluoride	PF_3	T,C	0.1
Silane	SiH_4	P,F	5
Silicon tetrafluoride	SiF_4	T,C	not listed

*T = Toxic, C = Corrosive, F = Flammable, P = Pyrophoric[7]

Chemically, dopant gases are hydrides of arsenic, boron, silicon, phosphorus, gemarium and antimony. The corresponding gases are arsine (AsH_3), diborane (B_2H_6), silane (SiH_4), phosphine (PH_3), germane (GeH_4) and stibine (SbH_3). All are either flammable or pyrophoric, and as a group have pungent odors and are extremely toxic.[8,9]

More recent legislation in the form of UFC Article 80 (revision) and the Toxic Gas Model Ordinance has expanded the definition and classification of toxic gases.

Article 80 defines two categories of toxic gases:

a. Toxic gas: A gas that has a median lethal concentration (LC_{50}) in air of 200-2000 ppm exposure for one hour or less, or a Class B Poison Gas. Examples: chlorine and hydrogen sulfide.

b. Highly toxic gas: A gas that has an LC_{50} of less than 200 ppm exposure for one hour or less, or a Class A Poison Gas. Examples: arsine, fluorine, cyanogen, hydrogen cyanide, nitric oxide, phosgene, phosphine.

The Toxic Gas Model Ordinance defines a toxic gas as any gas meeting the following criteria:

a. LC_{50} of 200 ppm for one hour or less, or LD_{50} (skin absorption) of less than 200 mg/kg body weight for 24 hours or less.
b. Poison A hazard class as defined by Code of Federal Regulations (CFR) 49.[10]
c. Extremely hazardous substance as defined by CFR 40.[11]
d. Determined through appraisal, testing, or objective means to be likely to create a significant hazard to public safety.
e. Determined by the Fire Chief to be likely to create a significant hazard to public safety.

STORAGE OF TOXIC GASES

Existing codes (Article 51 & UBC H-6) give two options for the storage of toxic gases: Exterior storage that complies with separation and fire safety requirements; or interior storage within an HPM storage (cut-off) room. Exterior storage of toxic gases must include the following:

a. Protection against physical hazards.
b. Security to prevent unauthorized access.
c. One-hour fire separation between toxics and other gases.
d. Sprinkler protection.
e. Emergency equipment (fire extinguishers, self-contained breathing apparatus).
f. Proper labeling and hazard identification.
g. Regular safety inspections.

Interior storage is permitted within an HPM storage room having at least one exterior wall at least 30 feet from the property line and not exceeding 6000 square feet. A two hour fire separation is required when the storage area exceeds 300 square feet. The maximum quantities of flammable and toxic gases (combined) must be less than 15,000 cubic feet.

Article 80 requires that interior storage of toxic and highly toxic gases is located in a "control area" conforming to UBC Group H, Division 7 occupancy requirements.[12] Ventilation requirements for gas cabinets and gas storage units specify that the exhaust system be capable of "processing the entire contents of the largest tank or cylinder of toxic gas stored".

These treatment systems must be designed to reduce the concentration of gas discharged to the atmosphere to one-half of the IDLH (immediately dangerous to life and health) which represents the maximum concentration level from which one could escape within 30 minutes without any escape-impairing symptoms or irreversible health effects.

A continuous gas detection system capable of monitoring the room at or below the PEL (permissible exposure limit) and at or below the IDLH at the discharge point of the treatment system is required. The PEL is the maximum permitted 8 hour time weighted average concentration of airborne contaminant. Exterior storage requirements specify distance limitations to exposure, fire protection, treatment systems and gas cabinets for leaking cylinders. The major difference between Article 80 and the Toxic Gas Model Ordinance as it applies to exterior storage is that the Ordinance requires secondary containment and a discharge treatment system to control an unauthorized release of toxic gas. The Ordinance also outlines specific requirements for new exterior and interior storage and use, as well as existing exterior and interior storage and use.

TOXIC GAS DISPENSING

UFC Article 51 and UBC H-6 make allowances for the use of toxic gases within a fabrication area. With few exceptions (lecture bottles of toxic gases within ion implanters for example), the enforcement trend is to require location of toxics exterior to the building piped into the wafer fab.

All of the previously mentioned codes have similar dispensing requirements for toxic gases. These include, but are not limited to:

a. Installation of toxic gases in gas cabinets of 12 gauge steel, equipped with self-closing, latching door and ventilation of not less than 200 ft per min at the face of access ports.
b. Approved fire sprinklers in gas cabinets.
c. Separation of toxic gases from other classifications with a limit of three cylinders per cabinet and proper labeling.
d. Excess flow control and automatic shut-off from a remote location.
e. Gas tubing must have welded connections unless located in an exhausted enclosure.
f. Continuous gas detection capable of shutting down gas cylinders at the source and initiating an alarm when permissible exposure levels are exceeded.
g. Exhaust from gas cabinets directed to treatment systems to prevent unauthorized release.

TOXIC GAS MONITORING

Continuous gas monitoring systems are required by Article 51 for toxic gases when used or dispensed and the physiological warning properties (odor threshold) for the gas are higher than the permissible exposure limit.[13] Article 51 also requires automatic shutdown on toxic gas lines and the initiation of an alarm to an emergency control station when the PEL is exceeded.

Article 80 is more specific and in addition requires that the detection system initiate a local alarm (visual and audible) and provide warning both inside and outside of storage areas. The monitoring system must be capable of monitoring the room or storage area below the PEL and the discharge from the treatment system at one-half the IDLH limit.

The Toxic Gas Model Ordinance requires monitoring systems to detect toxic gases when concentrations exceed the TLV-TWA (threshold limit value - time weighted average) for occupied areas and one-half the IDLH in unoccupied areas. The TLV-TWA is the concentration for a normal 8 hour workday, 40 hours per week that a normal worker may be repeatedly exposed to, without adverse effects. The ordinance also defines a CEL (community emergency level) which is the minimum toxic gas concentration for which protection or evacuation of the surrounding community is recommended.

DISPERSION MODELING REQUIREMENTS

One of the unique aspects of the Toxic Gas Model Ordinance is the requirements for a dispersion modeling test of toxic gases. This requires that facilities using toxic gases perform a dispersion modeling test for each toxic gas used at the facility and that the results be submitted to the Fire Chief for emergency planning purposes. If dispersion modeling indicates the concentration of a toxic gas exceeds the CEL level at any property line, additional requirements such as treatment systems or community warning systems may be required.

HAZARDOUS MATERIALS MANAGEMENT PLAN

Article 80 and the Toxic Gas Model Ordinance require Hazardous Materials Inventory Statements (HMIS) and a Hazardous Materials Management Plan (HMMP). An HMIS must be prepared for each building using hazardous materials and must include generic and common chemical names, chemical formulas, Department of Transportation identification numbers and maximum quantities stored.

The HMMP must incorporate the following:

a. Description of operations and processes.
b. General site and building floor plan designating hazardous materials storage.
c. Hazardous materials handling information.
d. Chemical compatibility and separation procedures.
e. A monitoring plan.
f. Security precautions.
g. Hazard labeling.
h. Inspection and record keeping.
i. Employee training.
j. Emergency response.

REFERENCES

1. Uniform Building Code, Ch 9, Sec 911, Gr H, Div 6 (1985).
2. Uniform Fire Code, Article 51 (1985).
3. Uniform Fire Code, Article 80 (revision) (1987).
4. Toxic Gas Model Ordinance: Final Report (1987).
5. R.A. Bolmen Jr., Semi. Int'l. Vol 9, No. 4, 156 (1986).
6. ACGIH, Threshold Limits Valves for Chemical Substances in Workroom Air (1987).
7. Semi-Gas Systems, Gas Reference Guide (1986).
8. R.A. Bolmen Jr., Semi. Int'l. Vol 9, No. 5, 231 (1986).
9. W. Baker, and A.L. Mossman, Matheson Gas-Data Book, 6th edition (1980).
10. Code of Federal Regulations, Title 49, Section 172.101 (1985).
11. Code of Federal Regulations, Title 40 (1985).
12. Uniform Building Code, Ch 9, Sec 911, Gr H, Div 7 (1985).
13. P. Burggraaf, Semi. Int'l. Vol 10, No. 11, 57 (1987).

TRENDS IN WORKER'S COMPENSATION IN THE UNITED STATES

Terry Clausen
Solar Energy Research Institute, Golden, CO 80401

INTRODUCTION

In this era of multimillion dollar jury verdicts, it seems useful to review when employees can sue employers for injuries and when they must pursue a remedy under a worker's compensation statute. This is particularly true as governmental safety regulations have increased dramatically both in number and complexity. This paper seeks to answer the question of whether or not employers' liabilities have been significantly increased. In order to be able to properly evaluate modern trends, it is useful to understand some of the history and also the unique legal character of worker's compensation insurance. I would like to specifically acknowledge at the outset two things: first, that this paper is in large part based upon Professor Arthur Larson's excellent treatise on "Workmen's Compensation Law". His treatise is so dominant in the field that it would be almost impossible to write on this subject without relying on his treatise. Secondly, to specifically acknowledge the tremendous number of women now in the work force, I have consciously not used the term workmen's compensation and have instead used worker's compensation.

A BRIEF HISTORY OF WORKER'S COMPENSATION

In early English and Germanic law, circa 1100-1200 A.D., a master was responsible for the death or injury of his servant if the servant died or was injured while on a mission for the master, or while being summoned to do the master's bidding. This responsibility was based upon what modern law would call "but-for" causation. The reasoning was simply if the master had not sent the servant on this particlar mission then he would still be alive (i.e., but for his master's orders he would still be alive). This general principle of a master's liability to his injured servants must have sufficiently addressed whatever social and legal problems arose for it remained basically unchanged for approximately 600 years. As the English common law developed, the principle that a master was generally responsible for his servants was replaced by the concept that the master was liable only if his negligence caused the injury. The law also developed the principle that the master could also be held to be vicariously liable for the wrongs committed by his servants. This second principle greatly expanded a master's potential liabilities, for the acts of his servants could now be imputed to the master. This did not seem to pose a significant problem in an agrarian society. However, the coming of the industrial revolution caused this to change.

The Growth of Industry and Reduction of Workers' Remedies

As industries grew there were significantly more accidental injuries than in the previously existing agriculturally based society. Industrialization was generally perceived as a great social good, and a broad master's liability was perceived as an impediment to industrialization. As a consequence, legal doctrines evolved which greatly reduced the legal remedies available to workers seeking redress for being injured in the work place. The major barrier was the common law itself which required the employee to show that the employer was negligent, and, in addition, other doctrines evolved which prevented an injured employee from recovering even if he could show the employer to be negligent or at fault. There were two doctrines which evolved at about the same time. The first to become widely adopted was the "fellow-servant rule"[1]. This rule basically held that an employer was not liable if an employee was injured as the result of the negligence of a fellow employee. This rule was quickly adopted in America[2]. The second major legal doctrine to evolve in this area was the doctrine of "assumption of the risk". This rule ignored all of the economic realities of the time and assumed that a worker who was familiar with the dangers of a certain job, and yet, voluntarily went to this dangerous place to work could not be heard to complain about injuries resulting from dangers to which he voluntarily exposed himself[3]. The final, and most significant, reduction in common law remedies available to injured employees came with the adoption of the doctrine of "contributory negligence". If the employee's actions contributed to his being injured in any way, no matter how slight his negligence might have been, then the employee could recover nothing from his employer for his injury and he was without a remedy[4]. To fully appreciate the impact of these legal changes one must examine data which classifies the causes of various types of industrial accidents by fault. The best data available happen to be old German data which cover from 4 to 20 million workers in a variety of industries over a twenty-year period[5]. The data are presented in Table 1 below.

Table 1
Classification of Causes of Accidents

(1)	Negligence or fault of employer	16.81%
(2)	Joint negligence of employer and injured employee	4.66
(3)	Negligence of fellow-servant	5.28
(4)	Acts of God	2.31
(5)	Fault or negligence of injured employee	28.89
(6)	Inevitable accidents connected with the employment	42.05

If an injured employee in order to recover has to prove that the employer's negligence was the sole cause of his injury, and we look at Table 1, it becomes apparent that the injured employee can only recover in 16.81% of the cases. To say it another way, under the English common law, which by and large became America's law, 83% of persons injured in industrial accidents had no remedy available to them. And still others would lose because of the assumption of the risk doctrine and other vagaries of the legal system.

It is clear that this system created a result which was intolerable in a civilized society. There were many people being permanently disabled, or even killed, and neither they nor their families had any recourse. At that time, the Anglo-American legal systems were totally driven by concepts of fault. Whoever was wronged sued, and if he could prove the wrong, then he could recover his damages. This system, while internally consistent, was not going to generate an answer for the problem at hand. For this we had to look outside of traditional Anglo-American legal principles.

The Evolution of Worker's Compensation

The first American responses to this bad situation were both legislative and judicial. They tried to soften the harsh effects of the common law doctrines, but offered no new approaches. While mitigating some of the effects of assumption of the risk and contributory negligence, they were still pinned on the basic common law idea that the employer was liable to the employee only if it was his negligence which caused the employee's injury[6]. Many studies were undertaken in an attempt to determine how the old system was treating injured employees. One conducted in Illinois showed that out of 614 death cases, the families received absolutely nothing in 214 of the cases, 289 of the cases were settled for the average of a few hundred dollars, and the remaining cases were pending litigation[7]. What was needed was a new social principle, and America found the example needed in Europe, particularly in Germany.

The New Principle

The origin of almost all of the worker's compensation statutes which exist today can be found in nineteenth century Prussia (later Germany). Although Bismark was firmly in control, this was nevertheless a period of great philosophical debate about the "nature of the State" and the proper functions and responsibilities of government. As a result of much public and scholarly debate, it came to be accepted among the more prominent philosophers and statesmen that many of the misfortunes of life which befell individuals were ultimately societal in origin, not the fault so much of the individual. Into this category they placed industrial accidents. Out of this conclusion evolved a well developed and often eloquently stated argument, that the only way to effectively and fairly deal with this important social issue was industrial insurance[8]. During this same period Bismark was becoming more and more concerned with the increasing electoral strength being shown by the Marxian socialists. Therefore, in 1881, Bismark proposed to the Reichstag a comprehensive plan for compulsory industrial insurance[9].

Rapid Acceptance in Europe

The first comprehensive worker's compensation statute was enacted by Germany on July 6, 1884. No other European country at that time had a formal worker's compensation statute; however, during this same time period, there had been in several countries modifications of the laws, which tended towards the principle that industries should pay compensation to workers who were injured or disabled in the course of their employment. Virtually all of the countries in Europe, plus Australia and Canada, had enacted worker's compensation statutes ten or more years before the first attempts were made to pass such laws in the United States[10].

Development in the United States

By the early 1900's, America was ready for radical change. There was increasing social and political concern over the increasing number of industrial accidents, and the decreasing remedies available to those who were injured. In 1893, a full account of the German system was written by John Brooks and published as the Fourth Special Report of the Commissioner of Labor. It was widely read by legislators throughout this country. Then the British passed the Compensation Act of 1897 installing worker's compensation throughout Britian. This act was later to become the model for many states' worker's compensation statutes[11].

As one might expect, the state legislatures moved forward cautiously and the courts reacted relying upon traditional and accepted legal doctrines. In 1902, the Maryland Legislature passed an act to provide an accident fund for workers' in the very dangerous mining industry. The law was held to be unconstitutional[12]. The Montana Legislature in 1909 passed an act to protect its miners. This act was also held to be unconstitutional[13]. So too was a more comprehensive New York act passed in 1910[14]. In our present age of pervasive and detailed governmental regulation, it is difficult to image the violent and emotional constitutional debates which were provoked by these first attempts at worker's compensation laws. Basically, the courts' holdings were based upon the rationale that to compel an employer to involuntarily pay into a fund without fault, or trial, was an unconstitutional taking of his property without the due process of law. To put this in historical perspective: at this time in American history an income tax was still unconstitutional.

The legislatures reacted to the court decisions and attempted to design statutes that would address the constitutional issues raised by the courts. Led by New Jersey, a number of states passed statutes which addressed the compulsory or involuntariness issue. They passed statutes which provided that employers could "elect" to be covered by worker's compensation. To provide incentive for the employers to elect, they provided in the same statutes that employers who did not elect were subject to common law suits, but without the benefit of the three popular common law defense discussed above[15]. A different tack taken by a number of state legislatures was to make the legislation less comprehensive and to limit compulsory coverage only to those industries know to be "hazardous". Both of these legislative efforts proved to be successful. Still there was continuing pressure to broaden the coverage to include more workers.

In 1917, the United States Supreme Court finally resolved the issue when it upheld the constitutionality of differing worker's compensation statutes from New York, Iowa, and Washington. The New York statute was compulsory and required almost all employers to participate. Once the rulings were handed down and the constitutional issues removed, the number of states adopting worker's compensation statutes expanded rapidly. By 1920, all but eight states had worker's compensation statutes in effect[16]. In the states where the coverage remained optional, most employers elected to be covered. In 1960, approximately 78% of the total American workforce was covered by worker's compensation. The primary exceptions were and remain domestic and agricultural workers[17].

THE UNIQUE NATURE OF WORKER'S COMPENSATION

Worker's compensation statutes introduced a new concept into the Anglo-American legal system because they are not based upon relative fault or merit. This section will illustrate that by contrasting worker's compensation with social insurance (the British system) and tort law (the traditional American way to right a wrong). Let us first look at the features of the typical worker's compensation statute.

- The employee is automatically entitled to certain benefits whenever the employee suffers a "personal injury by accident arising out of and in the course of employment."
- Negligence and fault are immaterial.
- Only employees are covered.
- Benefits payable are based on the wages being earned up to a statutory maximum and are typically one-half to two-thirds of wages, plus payment of hospital and medical expenses.
- In exchange for these modest but assured benefits, the employee gives up the right to sue the employer for damages.
- Benefits are only paid for injuries that produce disabilities that reduce one's ability to earn income.
- The right to sue third persons whose negligence may have caused the injury remains, but if one sues and wins the employee must then pay back to the employer any benefits received under worker's compensation.
- Employers are required to assure their ability to pay for workers' injuries by obtaining and maintaining private insurance or subscribing to a state run insurance fund.
- The premiums paid by the employer will be adjusted based on experience; i.e., if you have more injured employees you will have to pay higher premiums.
- The burden of worker's compensation costs is ultimately paid by the consumer of the product because it becomes a cost of production of that particular product[18].

By contrast if we were to use a system of social insurance then everyone who was injured would receive a flat rate payment, rather than having it tied to their salary. On the other side of the coin, all employers would pay the same rate regardless of their experience. Social insurance is administered by the government, while with worker's compensation employers can generally make private arrangements with insurers. Worker's compensation also serves the social good by allocating the liability to the products with the greatest hazards. The competitive disadvantage of higher insurance premiums (all other things being equal) could serve to drive the more hazardous product from the market. These are some of the differences between worker's compensation and social insurance. Let us now contrast worker's compensation with the area that is more important for this discussion: tort law.

Tort law is the area of the law that concerns itself with providing remedies to those who have been wronged, or injured, by the negligent or malicious behavior of others. It attempts to allocate blame, assign fault, and to make whole the person who was injured. Unlike tort law, worker's compensation does not concern itself with fault. There is only one question to be answered. Was the employee injured by accident while acting within the scope and course of employment? If the answer is in the affirmative then the employee is entitled to receive worker's compensation benefits. There are, however, limitations to worker's compensation. The injury must be the result of an accident which is an identifiable event. Occupational diseases are not covered because they result generally from long term exposure, not from an accident. Only injuries which reduce one's ability to earn a living are compensable. So it has been held that persons who were rendered impotent or unable to bear children as a result of accidents on the job could not receive any benefits under worker's compensation and neither could they sue their employer[19]. Neither can employees be compensated for, nor sue their employer for, pain and suffering, loss of taste, smell, or sensation, or disfigurement[20]. This should begin to make clear the major benefit of worker's compensation statutes to the employer: it becomes the <u>exclusive</u> remedy available to an injured employee. While the employer is responsible for almost all injuries to the employee on the job, they are paid for at the very modest statutory rates provided for by worker's compensation schedules, and the employer is not subject to suits in courts which might result in very large verdicts for the employee. In the beginning, worker's compensation statutes were probably a great disadvantage of the employer who, armed with all of the old common law defenses, could have prevailed in court and defeated most claims. Now the result is the opposite. The old common law defenses have long been abolished and juries are willing to bring in verdicts which they believe are necessary to truly compensate the victim, but which are shocking to others. It becomes clear very quickly to the injured employee, particularly one injured by the negligence or refusal of his employer to follow a safety rule, that if he can sue in court he stands to recover incredibly more than he will ever recover under worker's compensation. So this is where the more interesting and important legal battles are being fought today. The injured employee is seeking to escape from the exclusive remedy provided by worker's compensation, so that he can sue in court in front of a jury, and the employer is fighting to maintain worker's compensation as the only avenue available to an injured employee seeking redress. The remainder of this paper shall address this ongoing battle and will look briefly at a trend within existing worker's compensation statutes.

THE EXCLUSIVITY DOCTRINE

It is a fundamental principle of all worker's compensation statutes that payment under the statute is the exclusive method of recovery for injuries occurring on the job. The theory behind this is that the law provides employees with prompt and sure payments without any proof of negligence on the part of the employer. The employer, in return for accepting vicarious liability for all work related injuries (and thereby relinquishing all traditional defenses which would have been available), is shielded from tort litigation. The exclusive remedy provisions only apply if the injury itself falls within the coverage provisions of the worker's compensation statutes. There is generally a four-part test: the employee must (1) suffer a personal injury, (2) as the result of an accident, (3) which arose out of, and (4) in the course of employment. One should also note that it is the availability of the coverage that controls. If coverage is available, whether or not the plaintiff receives compensation, then worker's compensation provides the exclusive remedy. If an injury is not of the type covered by the statutes, then a suit is not barred; some examples would be:

- intentional injuries by the employer or supervisory personnel;
- actions for libel or slander;
- actions alleging only non-physical injuries which were not the result of an accident (more of these are being covered, see below);
- suits against insurers for intentional or outrageous failure to pay claims (barred in some states by a statutory penalty);
- actions for retaliatory discharge for filing a worker's compensation claim; and
- suits against third parties; most states now prohibit suits against co-employers for work-related injuries.

Intentional injury by the employer requires specific intent to injure. Intentional, gross, wilful, or malicious negligence is not sufficient. Neither is knowingly permitting a hazardous work condition to exist, wilfully and unlawfully violating a safety statute, or ordering an employee to do dangerous work[21]. In fact, OSHA contains an explicit provision that it does not in any way affect liabilities related to any worker's compensation laws. So, a violation of OSHA does not affect the exclusivity of the worker's compensation remedy[22]. Some states provide for a percentage increase in the worker's compensation, payable as a penalty for an employer's wilful misconduct or failure to provide safety devices[23]. Florida has held that violating an OSHA standard is sufficient to invoke the statutory penalty[24] while New Mexico has held that neither a state nor a federal OSHA violation is sufficient to trigger the 10% penalty provision in the New Mexico statute[25].

Let's review some cases to see how courts are interpreting the above general rules. In a California case, an employee was injured by a plastic molding press that was not equipped with the proper safety devices. The court held his suit barred by the exclusive remedy provisions of the worker's compensation law[26]. In Florida, the plaintiff's husband died as the result of unprotected exposure to chemicals being sprayed from a truck by his employer. The court held worker's compensation statute barred the suit[27]. In Idaho, a physical therapist injured his back and his doctor told him to take two weeks off to recover. His employer was aware of his injury, but threatened to fire him if he did not work. He reported to work as ordered and was injured. The court held his action barred by the exclusive remedy provision of the worker's compensation law[28]. In another case the plaintiff was raped while working in a private home for juvenile delinquents. She brought suit alleging that the boy who raped her was negligently admitted. The court while holding that being raped was not a compensable injury under the worker's compensation statute, also ruled that her injuries were sustained in the course of her employement and the alleged negligence occured during the course of the employment, so her suit was barred[29]. A Colorado employee injured his back in the course of his employment. He returned to work after a year and a half off. His doctor specifically recommended that he be assigned to light work. His employer assigned him to a job requiring heavy lifting and he injured his back again. The court dismissed his suit saying worker's compensation was his sole remedy[30]. In a case out of Connecticut, the court dismissed a suit which alleged violations of state and federal OSHA statutes and reaffirmed the rule that nothing short of deliberate intention to produce injury will remove the exclusiveness bar[31].

As can be seen from the above cases, the exclusiveness doctrine is alive and well and is being strictly interpreted and applied almost without exception. However, one exception has developed, although to date it has been rejected in all but two states (California and Ohio and one federal case).

The Dual Capacity Doctrine

Like most legal doctrines, the dual capacity doctrine had its origins in a unique set of facts which the court could not address to its satisfaction using existing rules. In this case the plaintiff was a nurse and an employee of a California chiropractor who injured herself while at work. Her employer treated her injuries, but the treatments were negligently administered and aggravated her earlier injury. The plaintiff received worker's compensation for her injury and then sued for the negligent treatment of her injury. The court had no problem reasoning that the chiropractor had two separate and distinct relationships with the plaintiff: one as her employer and the other as her doctor. There was a separate set of legal responsibilities which flowed from each relationship. As her employer, the defendant had no obligation to treat the plaintiff, but once he chose to do so he assumed an obligation, as any other doctor, to treat the patient properly[32]. Although the court did not use the phrase, reviewers of the case have concluded that the exclusive remedy provisions of worker's compensation did not apply, because the doctor was acting in a "dual capacity" as both her employer and her doctor.

In a later California case, an employee who was injured when some scaffolding on which he was working collapsed sued his employer on a products liability theory. The employer had manufactured the scaffolding. The court held that the employee had a cause of action for products liability, notwithstanding that the defendant was the employer, and the injury occurred in the course of employment, provided that the product which caused the injury was manufactured not just for the use of the employer but was sold to the public[33]. Here, the dual capacity is that of employer and manufacturer. These two principles have been written into the California labor code. However, other than one case in Ohio[34], all other courts have rejected the doctrine on one of two grounds: (1) that it could totally erode the exclusivity principle and throw worker's compensation into chaos, or (2) that the adoption of such an exception is best left to the legislature. Some commentators have urged the courts to apply the dual capacity doctrine even though it would make the court's job more difficult[35]. The exclusive remedy doctrine, while under attack, seems to be holding up well. We conclude with a brief look at a trend which is expanding the injuries covered by worker's compensation.

Expanding Acceptance of Mental Disorder and Stress Claims

With present day knowledge of medicine, it cannot be denied that one's mental condition can produce tangible effects on one's physical functioning. If this is so, then one can make a rationale argument that if disabling mental conditions arise as a result of employment, they should be considered compensable injuries. The fact that such injuries introduce difficult issues of proof ought not to preclude a whole class of injuries from consideration. The expanding acceptance of stress claims seems to be an identifiable trend.

Courts have long recognized that physical injury can be accompanied by emotional difficulties. Where one's psychological trauma then contributes to one's overall disability, courts have uniformly held the entire disability to be compensable[36]. Where mental stress during work hours has led to physical trauma, usually heart attacks, almost all courts have awarded compensation[37]. The more difficult cases are those of employees who become psychologically disabled without any apparent physical injury. On these cases, the courts have been split for years. A majority of courts now would probably award compensation[38].

Psychological injuries without accompanying physical injuries have been held compensable where workers were disabled by witnessing tragic events, receiving notice of employment termination, being subjected to harassment and abuse, and experiencing excessive employment stress[39]. In Lousiana, a salesman suffered a nervous breakdown after being transferred to another store, having his salary cut, and being under a lot of pressure from his supervisor. This was held to be a compensable injury under worker's compensation[40]. In a West Virginia case, the court allowed compensation where the claimant developed a depressive neurosis as a result of being "cursed, grabbed, and harassed" by her supervisor while at work[41]. Other cases have awarded compensation where work related stress or injuries led to the worker committing suicide[42].

SUMMARY

The exclusivity rule is holding firm, employees who are injured on the job are being required to seek compensation under worker's compensation statutes, and other lawsuits are being barred. Only two states to date have accepted the dual capacity doctrine.

On the other hand, recent decisions have established a clear trend for awarding compensation whenever a work related injury disables a claimant, whether the injury is mental or physical. So, we are likely to see more claims for stress related injuries. Overall worker's compensation, as it was designed and implemented 150 years ago, still seems to be successfully addressing most of the problems which have arisen in the area of employment related injuries.

REFERENCES

1. Priestly v. Fowler, 3 M. & W. 1, 150 Reprin 1030 (1847).
2. Farwell v. Bosten & Worcester R.R., 4 Metc. 49 (Mass. 1842).
3. Arthur Larson, Workmen's Compensation Law, Volume 1 § 4.30.
4. Larson, supra, § 4.30.
5. Bureau of Labor Bulletin (January 1908) as quoted in Larson, supra, § 4.30.
6. Larson, supra, § 4.40, 4.50.
7. Larson, supra, § 4.50.
8. Larson, supra, § 5.10.
9. Larson, supra, § 5.10.
10. Schneider, Schneider's Workmen's Compensation, 3rd Edition, 1941, § 9.
11. Larson, supra, § 5.20.
12. Franklin v. United Rys. & Elec. Co. of Baltimore, 2 Baltimore City Rep. 309 (1904).
13. Cunningham v. Northwestern Improvement Co., 44 Mont. 180, 119 p. 554 (1911).
14. Ives v. South Buffalo Ry, 201 N.Y. 271, 94 N.E. 431 (1911).
15. Larson, supra, § 5.20.
16. Larson, supra, § 5.20, 5.30.
17. Larson, supra, § 5.30.
18. Larson, supra, § 1.10.
19. Kline v. Arden H. Verner Co., 503 Pa. 251, 469 A. 2d 158 (1983) and Heidler v. Industrisal Comm., 14 Ariz. App. 280, 482 P. 2d 889 (1971).
20. Larson, supra, § 65.50.
21. Larson, supra, § 68.13.
22. Larson, supra, § 65.34.
23. Larson, supra, § 69.20.
24. McKenzie Tank Lines, Inc. v. McConley, 418 So. 2d 1177 (Fla. App 1982).
25. Casillas v. S.W.I.G., 96 N.M. 84, 628 P. 2d 329 (N.M. App) cert. den. 96 N.M. 116, 628 P. 2d 686 (1981).
26. Balido v. Improved Mech., Inc., 29 Cal. App 3d 633, 105 Col. Rptr. 890 (1973).
27. Revere v. Shell Chemical Inc., 376 So. 2d 1214 (Fla. App 1979).
28. Cope v. State, 108 Idaho 416, 700 P. 2d 38 (1985).
29. Rathbun v. Starr Commonwealth for Boys, 145 Mich. App. 303, 377 N.W. 2d 872 (1985).
30. Eason v. Frontier Airlines, 636 F. 2d 293 (10th Cir. 1981).
31. Mingachos v. CBS, Inc., 196 Conn. 91, 491 A. 2d 368 (1985).
32. Duprey v. Shane, 249 P. 2d 8 (1952).
33. Douglas v. E & J Gallo, 69 Cal. App. 3d 103, 137 Cal. Rptr. 797 (1977).
34. Mercer v. Uniroyal, 361 N.E. 2d 491 (1976).
35. 8 Thurgood Marshall L. Rev. 384, (Spring 1983).
36. 30 S.D.L. Rev. 636 (Summer 1985). p. 638, note 29.
37. Id, p. 639.
38. Larson, § 42.23.
39. 30 S.D.L. Rev. 636, 639 (Summer 1985).
40. Taquino v. Sears, Roebuck and Company, 438 So. 2d 625 (La. Ct. App. 1983).
41. Breeden v. Workmen's Compensation Commission, 385 S.E. 2d 398 (W. Va. 1981)
42. 30 S.D.L. Rev. 636, 643 (Summer 1985).

NATIONAL AND INTERNATIONAL CENTERS FOR ENVIRONMENTAL, HEALTH AND SAFETY INFORMATION

A.O. Bulawka
Photovoltaic Energy Technology Division
U.S. Department of Energy
Washington, D.C. 20585

ABSTRACT

In February 1987, the U.S. Department of Energy (USDOE) and the Biomedical and Environmental Assessment Division at Brookhaven National Laboratory hosted the First International Expert Working Group Meeting on Environmental, Health and Safety Aspects of Photovoltaic Energy Systems. This meeting was attended by representatives from France/Germany, Italy, Japan and the U.S. Over the course of the two day workshop, discussions focused on the goals, functional needs and responsibilities for national and international assistance centers and on exemplary research tasks. As an outgrowth of this Workshop, USDOE is now finalizing plans to establish a National Assistance Center.

INTRODUCTION

Materials and equipment used in photovoltaic cell manufacture may present risks to environment, health and safety (EH&S) because of their inherent biological (i.e., toxic or carcinogenic) and physical (e.g. radiofrequency radiation, electric shock) properties. The future commercial success of the photovoltaics industry will depend not only on the technical feasibility of producing photovoltaic devices at lower cost, but also on industry's ability to manage effectively associated EH&S hazards. It is important, therefore, that EH&S hazards be examined concurrently as the technology matures. Early identification and examination can provide system designers and decision-makers with important flexibility to develop timely and efficient mitigative strategies before large-scale commitment of time and resources. In this way, control costs and risk to health and environment can be minimized, while simultaneously maximizing public acceptance. In recognition of the importance of managing EH&S hazards as the technology matures, the U.S. Department of Energy (USDOE) and the Biomedical and Environmental Assessment Division at Brookhaven National Laboratory (BNL) hosted the First International Photovoltaic Energy Project (PEP) Expert Working Group Meeting on EH&S Aspects of Photovoltaic Energy Systems. As an outgrowth of this meeting, DOE is now finalizing plans to establish a National Assistance Center. This paper summarizes the discussions of the PEP meeting and the planned activities of the National Assistance Center.

PEP INTERNATIONAL EXPERT WORKING GROUP MEETING

PEP is an international organization of government agencies overseeing photovoltaic research programs. Member countries participating in this effort include the U.S., Japan, Italy, France, Germany, and England. At the 1985 PEP meeting in Nice, France, the Director of the USDOE Photovoltaic Energy Technology Division proposed that an International Expert Working Group on EH&S Aspects of Photovoltaic Energy Systems be convened. This proposal was enthusiastically endorsed by this governing body. As an outgrowth of this proposal, the Biomedical and Environmental Assessment Division (BEAD) at Brookhaven National Laboratory (BNL) and the USDOE hosted the First International Expert Working Group Meeting on Environmental, Health and Safety Aspects of Photovoltaic Energy Systems, February 24-25, 1987.[1] This two-day meeting was attended by representatives from France/Germany, Italy, Japan and the U.S.. At this meeting, each participating organization provided an overview of their organization's EH&S activities. Discussion then focused on the overall aims, objectives, and principles of implementation for an Expert Working Group.

As a result of these discussions, meeting participants concluded that international cooperation is needed to:

- (i) stimulate industry and governments to implement research programs designed to reduce or mitigate hazards which could hinder the technology's development and adversely affect EH&S;
- (ii) focus ongoing activities by governments and industry to assure that all important EH&S issues are evaluated;
- (iii) identify established governmental, quasi-governmental, or industry organizations engaged in EH&S research programs;
- (iv) encourage the establishment of assistance centers which will act as repositories for existing information and respond to information needs of governments and industry;
- (v) transfer information gained from activities;
- (vi) train specialists to assist government and industry; and,
- (vii) continue exchange among the international community on the status of photovoltaic EH&S issues.

In addition to these basic needs, seven EH&S research tasks which could contribute to the existing resource base were identified. Since there was little time at the Workshop to discuss these tasks, they are introduced below simply as a basis for future discussions by the overall photovoltaics community.

Task 1: Identification of EH&S Aspects of Materials and Technologies

Research during the last 10 years has resulted in many new technical approaches for making less costly and more efficient photovoltaic cells. Early technologies were improved and promising new ones are emerging. Concurrently photovoltaic firms are moving ahead with plans for large-scale commercialization of new technologies. These technologies use chemical substances and manufacturing processes not used before in large-scale production. Several of these materials are highly toxic and/or flammable. There are also physical hazards (e.g., electrical and electromagnetic fields) associated with some manufacturing systems. One must ensure that public and occupational health and safety are not endangered as new materials and production processes are explored. The purpose of this task is to develop early examinations of the EH&S risks in the laboratory environment where most of the activities currently take place, and in pilot plant and commercial-scale environments, to aid in choosing safer material/processes alternatives, reducing engineering control costs, and minimizing health risks.

Task 2: Compendium of Safe Practices

A compendium - perhaps as a loose-leaf book - of "Safe Practices" used for different material and process options would be useful for facilities engaged in photovoltaic cell (research and development) R&D or production. Individual pages could be updated or revised as necessary with current revision dates on each page. Incomplete sections would not restrict use of the completed sections. The compendium would contain the following kinds of information:

- (i) equipment descriptions;
- (ii) suppliers of equipment;
- (iii) facilities required;
- (iv) proper tools and personal safety equipment;
- (v) a narrative on extent of the hazard and any known safety precautions which might be overlooked;
- (vi) emergency procedures with as much quantitative information as is known (subject chemical volume and expected dilution with distance, effect of water and or fire, secondary chemical reactions to be avoided, heavier or lighter than air, etc.).

In this compendium, notation should be used to distinguish between "tried and true" safety procedures and those thought to be valid by way of logical extension of good practice. Current practices for managing toxic or hazardous materials or processes in the photovoltaics industry draw heavily on existing experiences from the semiconductor and chemical industries. This experience base is a useful source of data for the compendium. The most accessible source of information, however, is from suppliers of chemicals and chemical handling equipment. In the U.S., suppliers of hazardous gases have prepared information (i.e. material safety data sheets) about toxicity levels and handling procedures. Companies selling gas handling equipment, storage cabinets, exhaust venting hoods, fume scrubbers, contamination level instrumentation, etc., have technical consultants to assist in use of their products. Consulting engineering firms which design semiconductor facilities are sources of overall system design information. They are familiar with fire codes, building codes, and matters of building configuration such as evacuation problems, and explosion containment, not usually the concern of chemical suppliers or handling equipment suppliers.

Task 3: Assistance Centers

To provide EH&S information to researchers and plant managers in the photovoltaic industry, the formation of National Assistance Centers is suggested. These Centers could be located within some national organization associated with a centralized photovoltaic effort. Such Centers could be established in each of the participating countries to serve their respective constituencies or as a coordinated international facility. The Centers would keep abreast of the published literature in the field, provide relevant reprints and literature references to inquiries, and provide answers to inquiries from industry, universities, and other parties.

Centers could also provide access to experts, who would be available on request, to help universities or industry set up or improve their safety procedures, equipment, or organization. They could also conduct research into areas where information is lacking (e.g. into pollution-control equipment effectiveness).

National Assistance Centers should meet to share information and to identify areas which lack information. An International Assistance Center should compile results from the National Centers. Similarly, an International Assistance Center should provide direct support to industries, governments (e.g., developing countries) or international organizations (e.g., United Nations Development Program) in need of EH&S data.

Figure 1 depicts a proposed framework for the transfer of information through these organizations. As now proposed, the U.S. National Assistance Center to be implemented at BNL will act as the focal point for the International activities.

Task 4: Data Generation

EH&S information from photovoltaic R&D and manufacturing facilities is limited. Thus, the purpose of this task is to: (i) identify which data gathering activities could be included in the operation of current and planned facilities; and (ii) develop an implementation plan for gathering needed data.

Collection of new data will help EH&S specialists identify handling and disposal needs for hazardous or toxic materials/processes. Precise monitoring instrumentation and data acquisition hardware may have to be developed and calibrated. Likewise, experiments may be needed to gain the desired information successfully. Specific data generation methods to be used will depend on type of information desired and available resources. The success of this effort will depend on the cooperation of the funding organizations (e.g., USDOE), the National Assistance Centers, and contractors.

Task 5: Incident/Accident Analysis

The purpose of analyzing incidents and accidents in the photovoltaic industry is to identify problem areas relating to health and safety risks and their mitigation. The results of such analyses, based on actual incidents and accidents, should provide valuable feedback for defining R&D needs in process and control system designs, as well as for identifying improvements in operational procedures and work practices. Incident and accident analysis includes the following steps:

(i) Compile and review past events in photovoltaic R&D and production facilities.

(ii) Establish the nature of these events (e.g., on-site explosion, large-scale leakage of toxic gases to the environment).

(iii) Identify and classify the causes of these events (e.g., human error, system failure, lack of maintenance).

(iv) Review existing control technologies and recommend mitigation options to prevent future events.

(v) Suggest R&D focus if current technologies or information are inadequate for effective risk reduction.

Because of the relative immaturity of the photovoltaics industry, historical data on events in individual countries are generally sparse. International cooperation to exchange and share information in this area can be beneficial in providing a larger data base in this developing industry.

Task 6: Guidelines and Standards

There are few EH&S guidelines or standards directed to the photovoltaic industry. Some hazards from materials and equipment are, however, regulated in other industries. Thus, the aim of this task is to identify and compile existing EH&S guidelines and standards applicable to R&D and commercial-scale facilities in each country. These could be included in the compendium described in Task 2 or published as a separate handbook. These compilations could also contribute to identification or establishment of "safe practices."

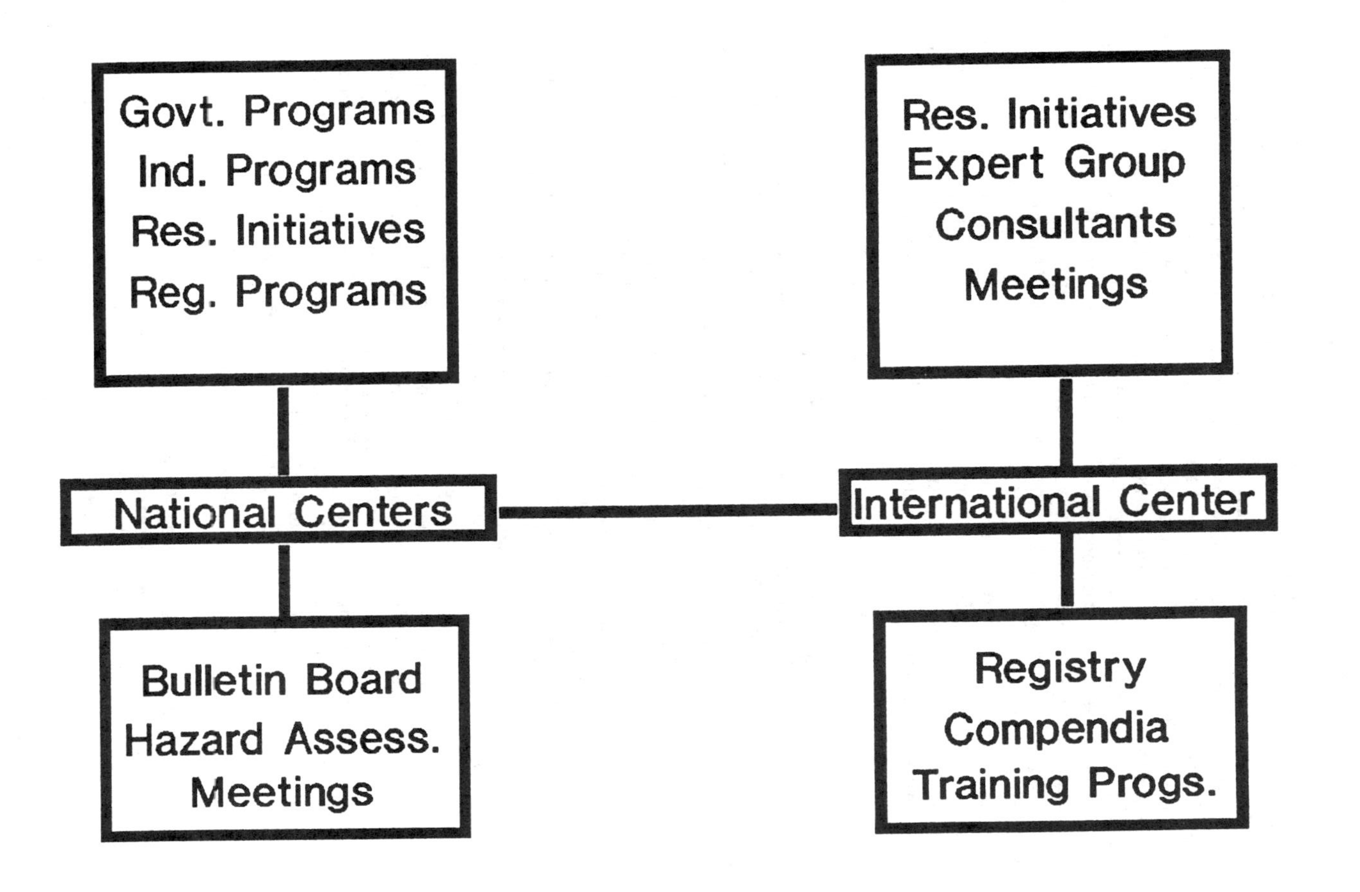
Govt. Programs
Ind. Programs
Res. Initiatives
Reg. Programs
National Centers
Bulletin Board
Hazard Assess.
Meetings
Res. Initiatives
Expert Group
Consultants
Meetings
International Center
Registry
Compendia
Training Progs.

Task 7: Worker Registry

The purpose of establishing a worker registry in the photovoltaic industry is to provide a statistical basis for the evaluation of the industry's occupational hazards. The data bank can serve as a source of information on the nature, extent, and cause of occupational illnesses and injuries occurring throughout the manufacturing process. The registry could also provide the basis for performing future studies to determine the presence of any health effects which may be associated with particular jobs or working conditions or to determine the effectiveness of exposure controls and work practices in reducing the incidence of observed illnesses or injuries. For this effort, latent health effects (e.g. carcinogenic) represent a particularly difficult problem since these symptoms usually develop many years after the initial exposure. The proposed worker registry forms a basis for the identification of these long-term effects and their effective control.

In view of the fact that individual country statistics on work related illness and injury are generally limited in number for this evolving industry, international cooperation to share each country's experiences could be crucial to the formation of a useful worker registry at this early stage of the industry's development.

NATIONAL ASSISTANCE CENTER

As a follow-up to the PEP Expert Working, plans are now being finalized to establish a USDOE funded National Assistance Center at BNL. The objective of this Center will be to provide focused EH&S information to DOE contractors and to the U.S. photovoltaic community. Consistent with this objective, tentative USDOE supported activities planned for the Center in FY 1988 include:

(i) Installation, operation and maintenance of a Photovoltaic EH&S Electronic Bulletin Board System (BBS) to foster increased communications and technology transfer among organizations and individuals engaged in the research, development and commercialization of photovoltaic energy systems. The BBS as now envisaged will act as a medium for the exchange of messages, bulletins, and data files. In addition, on-going conferences or guilds will be established to encourage detailed discussion on selected topics. USDOE hopes that the BBS will facilitate networking and permit users to query and rapidly respond to problems and issues of special or general interest, serve as a repository for these responses and provide updates about meetings, new citations, regulations etc. The BBS will be accessible to commercially and federally funded organizations engaged in photovoltaic cell research, development and production.

(ii) Initiation of on-site visits at USDOE contractor facilities to discuss specific problems and needs; and

(iii) Preparation of a Compendium of Safe Practices (see task 7 above).

CONCLUSIONS

In response to these ongoing discussions, USDOE has:

(i) Focused some research efforts at BNL to respond to some of the research tasks outlined;

(ii) Agreed to participate in the 2nd PEP Expert Working Group Meeting to be held in the Spring/Summer 1988;

(iii) Initiated efforts to establish a National Assistance Center at BNL.

USDOE is committed to advancing photovoltaic energy systems which are both commercially competitive and environmentally acceptable. These objectives can be met through the active implementation of the international and national activities described above, and through open communication and cooperation of the private sector.

REFERENCES

1. International Expert Working Group on Environmental, Health and Safety Aspects of Photovoltaic Energy Systems: A Workshop Report, February 24-25, 1987, Brookhaven National Laboratory, Upton, NY 11973.

APPLICATIONS OF FAULT TREE ANALYSIS TO THE DESIGN PROCESS *

R. W. Youngblood
Brookhaven National Laboratory, Upton, New York 11973

ABSTRACT

Fault tree analysis of a system can provide a complete characterization of system failure modes, i.e., what combinations of component failures can give rise to system failure. This can be applied to the design process at several levels: (1) confirmatory analysis, in which a fault tree development is used to verify design adequacy, (2) importance analysis, in which fault tree analysis is used to highlight system vulnerabilities, and (3) design optimization, in which fault tree analysis is used to pick the least expensive configuration from a collection of possibilities satisfying a given constraint. Experience shows that the complexity of real systems warrants the systematic and structured development of fault trees for systems whose failure can have severe consequences.

INTRODUCTION

This paper discusses applications of fault tree analysis (FTA) to the design process. Certain applications of FTA have been discussed in this connection for many years, and these will be summarized briefly. More recently, Hulme[1], Worrell, and coworkers [2,3,4] showed how to perform certain kinds of optimization within an FTA. This makes it possible to use FTA directly to select, from a system design containing multiple hardware options, a combination of hardware which (a) satisfies logically formulated constraints which can be reliability goals, and (b) is the cheapest combination satisfying (a). This will also be discussed.

Numerous large-scale risk assessments have been performed. These have involved extensive analysis of diverse scenarios, construction of large-scale event tree and fault tree models, and sophisticated quantification techniques. (See Refs. 5, 6, and 7 for applications to nuclear power plants.) The results of these analyses have been used in a range of high-level decisionmaking contexts, including design evaluation. The scope of this paper is much narrower: it is to show what information can be obtained from fault tree analysis of systems or combinations of systems at the design stage, and how designers can use this information. The paper is intended for system designers who are unfamiliar with fault tree analysis.

FAULT TREE ANALYSIS

To facilitate discussion of the uses of FTA, we must first state briefly what a Boolean expression is. The process of fault tree development will not be discussed here, but proper development of the fault tree is extremely important.[8] A great

---

* This work is based on studies performed under the auspices of the U. S. Nuclear Regulatory Commission.

deal has been written about fault tree development, especially from within the chemical process industry; formal approaches (e. g., digraph modelling) to fault tree development are especially worth exploring. Numerous references can be found in the volumes edited by Lees.[9] It should also be understood that certain shortcuts are taken in this presentation, in the interest of brevity.

A fault tree is a graphical representation of a logical expression which describes the ways in which a complex event can occur. Suppose that the complex event we wish to analyze is "Failure of System X to perform its function". Suppose further that System X contains two trains of hardware, and that successful operation of EITHER train means that the system succeeds. Then "System X fails" is equivalent to ["Train 1 fails" AND "Train 2 fails"]. If Train 1 contains components A, B, C, ... in series, then "Train 1 fails" is equivalent to ["Component A fails" OR "Component B fails" OR ...]. Proceeding in this way, one can develop a complete description of a complex event (system failure) in terms of simpler events (e.g., train failures), which are themselves described in terms of even simpler events (e.g., component failures), and so on. When the analysis reaches a level at which the events are as simple as is meaningful to analyze, one stops, and these events are "primary" events for purposes of the analysis. Typically, this description is represented pictorially by logic gates (See Fig. 1 for an example). All events on the fault tree are to be thought of as simple declarative sentences which are either true or false, and the gates on the fault tree specify the logical relationship between the input events and the output event. "Train 1 fails" is either true or false; "System X fails" is true if both "Train 1 fails" AND "Train 2 fails" are true, and "System X fails" is false if either train succeeds. Once all of this is established, the tools of formal logic can be applied to analyze the model.

If we associate a Boolean equation with each gate, then, by successive substitutions, we can arrive at a Boolean equation for the top event in terms of the primary events. This can be transformed into a very useful representation having the following properties. The form of the top event expression is a logical "OR" of individual terms which represent system failure modes. Each term is a logical "AND" of primary events; for example (see Fig. 1), one term might be ["Flow control valve 1 sticks shut" AND "Flow control valve 2 sticks shut"]. In order for the top event to occur, one or more of the terms must be true; and in order for any term to be true, all of its constituent primary events must be true. For "coherent" problems, these terms are called the minimal cut sets. If we complement the expression for system failure - that is, if we apply a logical NOT operation to the top event "System X fails" - then we obtain a Boolean expression for system success, in which each term, or success path, is a conjunction of failures which do NOT occur. For the above example, since Train 1 is capable of achieving system success, one success path is [NOT (Component A fails)] AND [NOT (Component B fails)] AND [NOT (Component C fails)] ... .

The Boolean expression for system failure contains all the system failure modes, defined in terms of primary events. As such, it is an interesting object in its own right; it is an exhaustive catalog of the ways in which the top event can occur. Additionally, it is a basis for estimating how probable the top event is, and assessing the significance of various combinations of primary events. For example, the probability of occurrence of an individual term (minimal cut set, or system failure mode) can be calculated by standard methods in terms of primary

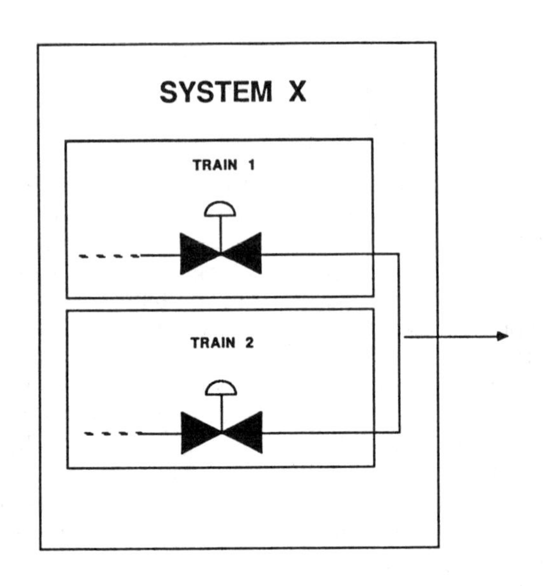

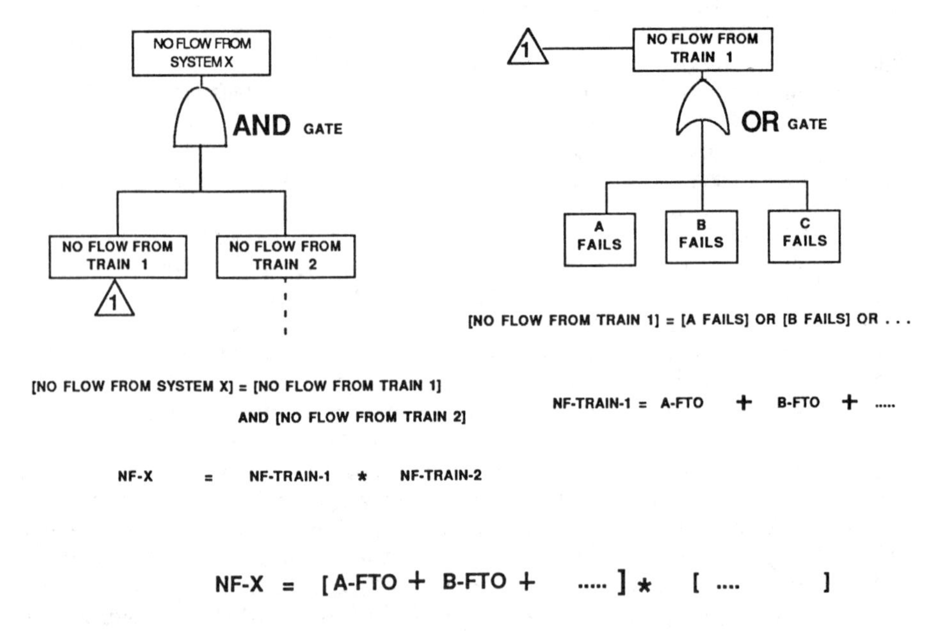

NF-X = [A-FTO + B-FTO + ...] * [....]

FIGURE 1. FAULT TREES AND BOOLEAN EQUATIONS

event probabilities (which may need to be conditional, if the events are not statistically independent). Similarly, these can be combined to yield estimates of top event probability. In practice, simple time-independent probabilities are frequently assigned to primary events, and the assumption is made that they are independent of each other. Term probabilities are then calculated as the product of the individual event probabilities. If it is also true that the term probabilities are small (say, much less than 0.1), then a very simple and useful approximation to top event probability is given by the sum of the term probabilities. Much has also been written about improvements on these approximations, but in many applications, they are not significant, with one key exception: the assumption that events are independent. If nominally independent events are actually linked by a common cause, this can have serious implications for system reliability. FTA tells the designer that a particular combination of events will fail the system; it cannot tell the designer that a single underlying cause can bring this about, unless this fact has been built into the model. Much has also been written on analysis of common cause failures, and while it continues to be controversial, numerous lessons can be learned from operating experience about common cause failures, what kinds of hardware are vulnerable to them, etc. These insights are not products of FTA, but are inputs to it, except that FTA can alert the designer to areas of the design in which common cause failure is relatively likely, potentially damaging, or both.

LOGISTICS OF FAULT TREE ANALYSIS

For realistic problems, fault trees must be processed by computer. There are many programs for doing this, with diverse capabilities. Many of them are listed in the Probabilistic Risk Assessment Procedures Guide[5], but many recent software developments are not listed there, especially those that are written for small computers.

Before any particular program is chosen, detailed thought should be given to how the results will be used. Some of the analyses discussed here can be performed with many programs, and some with few programs. The most flexible and powerful tool known to the present author is SETS [10,11], but even this program must be supplemented with other programs in order to perform all the computations mentioned here. For example, the optimization scheme discussed here can be performed only with SETS, as far as the present author is aware; some other fault tree programs have built-in capabilities for importance analysis, while SETS must be supplemented with other programs.

The availability of computational resources is another factor in choosing an approach. However, most of the real work in FTA is organizing the large volume of information which is encoded in the tree, and it is a poor allocation of resources to limit the usefulness of a labor-intensive fault tree development by settling for a superficial computer analysis.

BENEFITS OF FAULT TREE ANALYSIS

In simple situations, a fault tree analysis may appear to be no more than a graphical presentation of the obvious. What, then, is the point of developing a fault tree?

First of all, even in seemingly simple situations, there are real benefits to a systematic and structured approach to the assessment of designs. The process of breaking down the problem into small pieces, and consciously focusing on each piece, can lead to important insights. The point here is not that formal logic is needed to express these insights, especially with hindsight; rather, it has been found that a structured and systematic process of inquiry is much more productive than an unstructured approach. Testimonial to this point can readily be found in the wide literature surrounding HAZOP[12,13]. There is even a school of thought which holds that the insights achieved during the development of the model are the major benefit of fault tree analysis, and are more important than the final products of the analysis.

Second, realistic design issues can easily become sufficiently complex that a formal process of analysis becomes necessary in order to assure that important system vulnerabilities are not being overlooked. A real example of this is summarized below. This example, and others like it, demonstrate that the final products of the analysis can be extremely important. A key point of this example is that the system vulnerability discussed in this example was not spotted during development of the model, but rather emerged as an output.

Finally, in theory at least, fault tree analysis can be applied directly to design optimization. The literature on this point is relatively sparse, presumably because this approach has not yet been widely used. Therefore, demonstrations of the general practicality of this approach are so far lacking; as more designers explore these methods, consensus will develop regarding their usefulness.

DESIGN ASSESSMENT

The simplest way to use FTA in the design process is simply to decide whether the design meets some kind of reliability goal. If quantitative FTA is performed, then system failure probability can be estimated and compared with a numerical reliability goal; if a qualitative requirement has been formulated (e.g., that the system be able to perform its mission notwithstanding the occurrence of any single component failure), then the system failure modes can be examined individually to see whether the system design is adequate. If competing design alternatives are under consideration, an FTA of each can contribute to the selection process. This could involve cost-benefit analysis, or simple ranking of the estimated reliability of different designs.

Figure 3 is a simplified diagram of the electrical power system at Indian Point 3, which is a commercial nuclear power station located on the Hudson River in New York. Some of the electrical loads are also indicated on this figure; an important part of the Emergency Core Cooling System (ECCS) is the low pressure injection system, consisting in part of the two pumps labelled RHR 31 and RHR 32. These pumps derive motive AC power from nominally independent sources and are controlled by DC power derived from nominally independent busses. The ECCS is

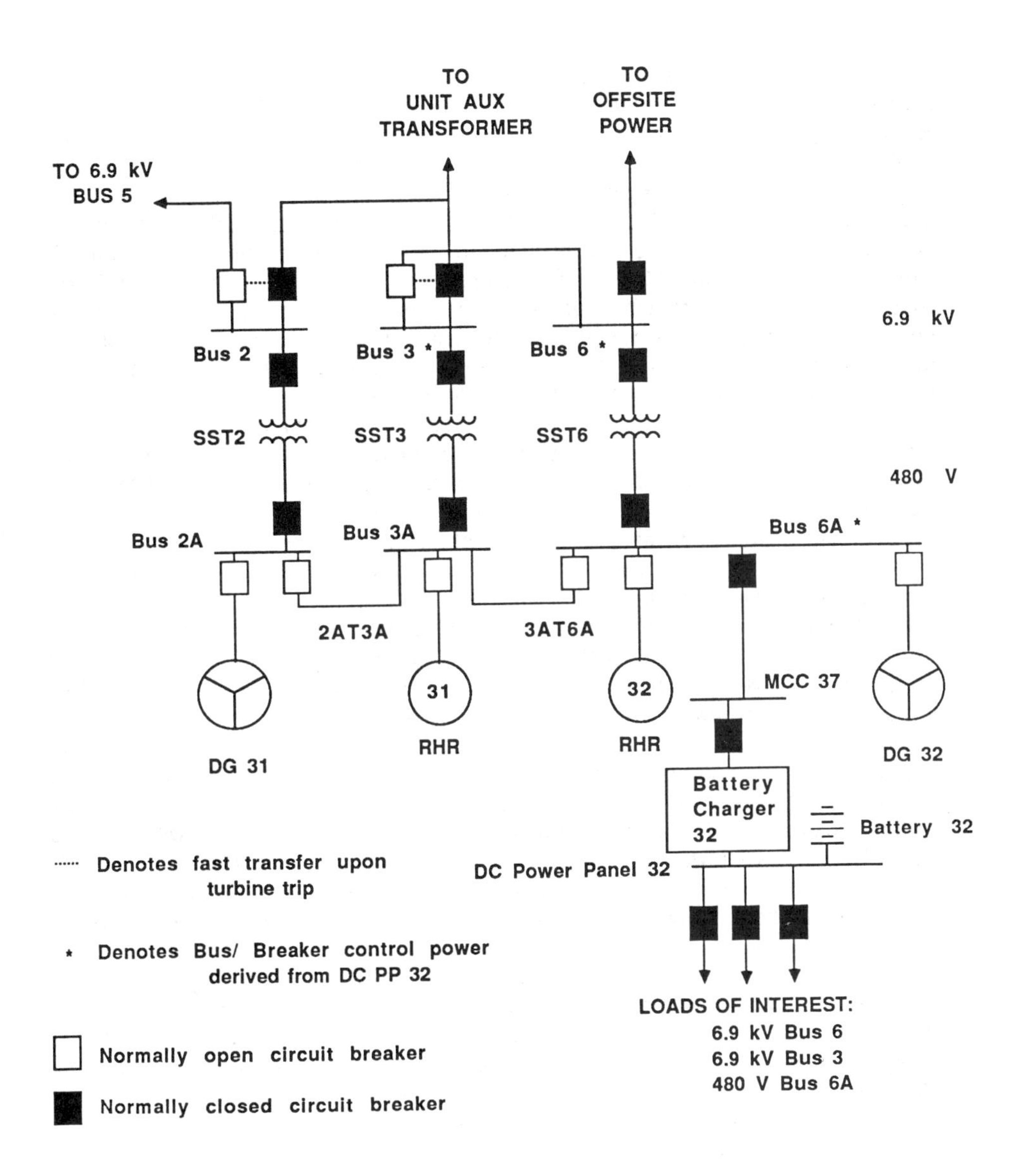

FIGURE 2. INDIAN POINT UNIT 3
ELECTRICAL POWER SYSTEM
NORMAL CONFIGURATION

legally required to possess sufficient hardware redundancy that no single component failure can cause loss of its function (the "single-failure criterion"). However, fault tree analysis of this system showed that lack of power from a single station battery (Battery 32) would cause system failure, even if electrical power from the offsite grid were available. Details can be found in Ref. 14; some of the pertinent highlights are the following. When a need for the ECCS is sensed, the actuation signal begins a series of automatic electrical load shedding operations which remove DC Bus 32 from its normal power source, forcing it to rely on Battery 32 (which is presumed dead, for present purposes). With DC Bus 32 now dead, certain other automatic switching operations are now stalled, including the starting of RHR 32 and the bus transfer of Bus 3 from its normal power source, the Unit Auxiliary Transformer, to offsite power from Bus 6. This deprives RHR 31 of motive AC power, because although backup power could be made available from Bus 2A, and would automatically be provided by diesel generators in a blackout, the present scenario is not enough of a blackout to cause the automatic re-energizing of Bus 3A. Thus, neither RHR pump functions.

This failure mode eluded the designers, the licensing review, and several other studies; it was found by a formal FTA, which was undertaken for purposes of methods development several years after the plant was licensed and operating. A team of analysts was involved; one person developed the fault tree model of AC distribution, another the DC system, and a third modelled the RHR pumps. The three fault trees were linked and analyzed by computer, and the finding emerged as a considerable surprise. It is noteworthy that a two-train fluid system was complex enough that formal FTA was necessary in order to assess compliance with the single-failure criterion, and that the finding was unexpected even by the individuals who prepared the system fault trees. More accurately, the complexity of this problem did not lie in the two-train fluid system, but in the switching and controls associated with a relatively complex electrical power system, whose unsuspected linkages compromised the nominal redundancy of the ECCS. There are many combinations of circumstances to analyze, and this particular combination had escaped attention. One further point can be derived from this instance: a complete loss of offsite AC power would have the effect of rescuing the ECCS, because AC power is automatically restored to Bus 3A. Loss of offsite power is not the worst case for this design, although it would be natural to assume otherwise; thus, simplified "worst case" analyses cannot be trusted, unless it is absolutely clear what the "worst" case is.

This episode demonstrates that a realistic level of complexity in system design makes it difficult to spot vulnerabilities in an *ad hoc* review process, and makes it almost impossible to have confidence in anything short of a careful FTA.

DESIGN IMPROVEMENT

Understanding the relative significance of different failure modes can help to establish priorities in the process of improving a design. Quantitative figures of merit called "importance measures" have been developed for this purpose. One widely used importance measure for primary events is the Fussell-Vesely (F-V) measure. (See Ref. 15 for a discussion of several importance measures.) The F-V importance of event X can be obtained by computing the probability of the union of all terms in the top event Boolean expression which contain event X, and dividing the result by the probability of the top event. Colloquially put, the F-V importance of an event is the fraction of top event occurrences which involve the event; clearly, then, reducing the probability of an event with a large F-V importance will lead to a significant reduction in failure probability. From this point of view, a listing of events ranked by F-V importance tells a designer where to look, if system reliability needs to be improved.

Another importance measure is the Birnbaum measure, which is essentially the derivative of top event probability with respect to the probability of event X; it is a measure of the sensitivity of top event probability to event X. It leads to a ranking of events rather different from that of the F-V measure; an event with a high Birnbaum importance does not necessarily help to account for a very high fraction of the top event probability, but would contribute substantially if its probability increased. Thus, a listing of events ranked by Birnbaum importance tells a designer what primary events must be defended against, lest they become significant contributors.

There are many other importance measures, but these two are illustrative and useful. The book by Henley and Kumamoto[15] includes a comparison of these and other single-event importance measures. One can also define importance measures for collections of events[16]; for example, one can ask what fraction of top event probability involves the coincidence of event X and event Y. The F-V importance of this coincidence can then be considered by analogy to the F-V importance of X. As noted above, pairs of events are typically asumed to be independent, but this need not be the case; if event X is a maintenance act and event Y is a random failure, and the coincidence of X and Y has a high importance, then measures can be taken by management to reduce or eliminate the likelihood of the coincidence, which may be easier than reducing either of the individual probabilities. Pair importances can also be used to assess the sensitivity of the FTA to the assumption that primary events are independent.

DESIGN OPTIMIZATION

Optimization and linear programming have been discussed for many years, as has FTA, but the demonstration by Hulme and coworkers[1-4, 17] of design optimization using fault tree software is fairly recent. The possibilities are best illustrated by two examples from their work. For historical reasons and for the sake of exposition, both are related to fire protection, but the ideas are more widely applicable.

1. Fire Protection [17]

Suppose that a system contains components in several different fire zones, and it is necessary to design fire protection (sprinklers, alarms, ...) in such a way that at least one success path is completely protected from beginning to end; to achieve this, multiple zones will have to be protected. In a complex situation, there may be many different ways of achieving this, some ways much more expensive than others. It is possible, using SETS, to catalog the ways and to select the cheapest. To do this, one prepares a special system fault tree whose logic reflects the locations of the vulnerable components, and whose minimal cut sets correspond to combinations of locations at which a fire would degrade any components that were located there. One also associates, with each zone, a cost of protecting it from fire. Then the path sets are collections of zones whose protection satisfies the design objective, and the cost of protection is the sum of the costs associated with each zone. Given the fault tree and the cost information, SETS can choose that collection of zones which contains at least one path set, and is the cheapest of all such collections.

2. Extra Fire Protection [1-4]

The previous example only addressed protection of a single path set; it showed how to provide a single layer of protection. In many real applications, however, a single layer of protection is not adequate; for example, a single layer of protection does not meet the "single-failure criterion" mentioned above. Suppose that we wish to meet the single-failure criterion for fires; that is, we wish to guarantee that no single failure in fire protection can cause loss of system function. The cheapest-path-set approach is not adequate for this, but it can be accomplished by the following steps.

For each cut set, note the pairs of zones whose protection satisfies the design objective; for example, if the cut set is "conflagration in zone A AND conflagration in zone B AND conflagration in zone C", then two layers of protection can be provided by [(protection of zone A AND protection of zone B) OR (protection of zone A AND protection of zone C) OR (protection of zone B AND protection of zone C)]. This is a requirement that must be met by any protection scheme purporting to offer two layers of protection, and it follows from the existence of the system failure mode A * B * C. Each cut set gives rise to a similar requirement, and the logical AND of these individual requirements gives rise to a logical expression whose terms correspond to different ways of allocating protection so as to meet the overall system requirement. This expression can be fed into SETS, which, as before, can manipulate it and choose the cheapest term.

SUMMARY

Fault tree analysis has long been established as a tool for confirmatory analysis, of the type described above under "Design Assessment". More recently, importance measures have attracted attention as tools for thinking about improvements to the design of systems and improvements to operational practices. The new techniques for design optimization are promising. All of these applications of fault tree analysis merit the consideration of persons who are designing systems which are complex and important. Experience has shown that the analysis of complex systems warrants a systematic, structured, and machine-aided approach, especially if the consequences of system failure are severe.

REFERENCES

1. Bernie L. Hulme and Richard B. Worrell, "Combinatorial Optimization with Boolean Constraints", SAND83-0085 (Sandia Report, Sandia National Laboratories, Albuquerque, New Mexico, February 1983).
2. Bernie L. Hulme, Lorraine S. Baca, Ann W. Shiver, Richard B. Worrell, "Boolean Computation of Optimum Hitting Sets", SAND84-0448 (Sandia Report, Sandia National Laboratories, Albuquerque, New Mexico, April 1984).
3. Bernie L. Hulme and Lorraine S. Baca, "A Boolean Approach to Zero-One Linear Programming", SAND84-1409 (Sandia Report, Sandia National Laboratories, Albuquerque, New Mexico, July 1984).
4. Bernie L. Hulme and Lorraine S. Baca, "Set Covering, Partition, and Packing", SAND84-0447 (Sandia Report, Sandia National Laboratories, Albuquerque, New Mexico, March 1984).
5. "PRA Procedures Guide", prepared under the auspices of the American Nuclear Society and the Institute of Electrical and Electronics Engineers, NUREG/CR-2300 (USNRC, Washington, DC, 1983).
6. D. Carlson *et al.*, "Interim Reliability Evaluation Program Procedures Guide", NUREG/CR-2728 (USNRC, Washington, DC,1982).
7. R.A.Bari *et al.*, "Probabilistic Safety Assessment Procedures Guide", NUREG/CR-2815 (USNRC, Washington, DC, 1985).
8. W. E. Vesely *et al.* , "Fault Tree Handbook", NUREG-0492 (USNRC, Washington, DC,1981).
9. F. P. Lees, ed. , Loss Prevention in the Process Industries (Butterworths, London, England, 1980).
10. R. B. Worrell, "SETS Reference Manual", NUREG/CR-4213 (SAND83-2675) (USNRC, Washington, DC, 1985).
11. Desmond W. Stack, "A SETS User's Manual for Accident Sequence Analysis", NUREG/CR-3547 (SAND83-2238) (USNRC, Washington, DC, 1984).
12. H. G. Lawley, "Operability Studies and Hazard Analysis", Loss Prevention 8, p. 105 (1974).
13. Chemical Industries Association, "A Guide to Hazard and Operability Studies", London, 1977.
14. R. Youngblood *et al.*, "Fault Tree Application to the Study of Systems Interactions at Indian Point 3", NUREG/CR-4207 (USNRC, Washington, DC, 1986).
15. Ernest J. Henley and Hiromitsu Kumamoto, Reliability Engineering and Risk Assessment (Prentice-Hall, Inc., Englewood Cliffs, N. J., 1981).
16. R. Youngblood, D. Xue, and N. Cho, "Pair Importance Measures in Systems Analysis", Proceedings of the International ANS/ENS Meeting on Thermal Reactor Safety, Feb. 2-6, San Diego, California (American Nuclear Society, LaGrange Park, Illinois, 1986).
17. Bernie L. Hulme, A. W. Shiver, and P. J. Slater, "Computing Minimum Cost Fire Protection", SAND82-0809 (Sandia Report, Sandia National Laboratories, Albuquerque, New Mexico, June 1982).

AUTHOR INDEX

AIP Conference Proceedings

		L.C. Number	ISBN
No. 1	Feedback and Dynamic Control of Plasmas – 1970	70-141596	0-88318-100-2
No. 2	Particles and Fields – 1971 (Rochester)	71-184662	0-88318-101-0
No. 3	Thermal Expansion – 1971 (Corning)	72-76970	0-88318-102-9
No. 4	Superconductivity in *d*- and *f*-Band Metals (Rochester, 1971)	74-18879	0-88318-103-7
No. 5	Magnetism and Magnetic Materials – 1971 (2 parts) (Chicago)	59-2468	0-88318-104-5
No. 6	Particle Physics (Irvine, 1971)	72-81239	0-88318-105-3
No. 7	Exploring the History of Nuclear Physics – 1972	72-81883	0-88318-106-1
No. 8	Experimental Meson Spectroscopy –1972	72-88226	0-88318-107-X
No. 9	Cyclotrons – 1972 (Vancouver)	72-92798	0-88318-108-8
No. 10	Magnetism and Magnetic Materials – 1972	72-623469	0-88318-109-6
No. 11	Transport Phenomena – 1973 (Brown University Conference)	73-80682	0-88318-110-X
No. 12	Experiments on High Energy Particle Collisions – 1973 (Vanderbilt Conference)	73-81705	0-88318-111–8
No. 13	π-π Scattering – 1973 (Tallahassee Conference)	73-81704	0-88318-112-6
No. 14	Particles and Fields – 1973 (APS/DPF Berkeley)	73-91923	0-88318-113-4
No. 15	High Energy Collisions – 1973 (Stony Brook)	73-92324	0-88318-114-2
No. 16	Causality and Physical Theories (Wayne State University, 1973)	73-93420	0-88318-115-0
No. 17	Thermal Expansion – 1973 (Lake of the Ozarks)	73-94415	0-88318-116-9
No. 18	Magnetism and Magnetic Materials – 1973 (2 parts) (Boston)	59-2468	0-88318-117-7
No. 19	Physics and the Energy Problem – 1974 (APS Chicago)	73-94416	0-88318-118-5
No. 20	Tetrahedrally Bonded Amorphous Semiconductors (Yorktown Heights, 1974)	74-80145	0-88318-119-3
No. 21	Experimental Meson Spectroscopy – 1974 (Boston)	74-82628	0-88318-120-7
No. 22	Neutrinos – 1974 (Philadelphia)	74-82413	0-88318-121-5
No. 23	Particles and Fields – 1974 (APS/DPF Williamsburg)	74-27575	0-88318-122-3
No. 24	Magnetism and Magnetic Materials – 1974 (20th Annual Conference, San Francisco)	75-2647	0-88318-123-1
No. 25	Efficient Use of Energy (The APS Studies on the Technical Aspects of the More Efficient Use of Energy)	75-18227	0-88318-124-X

No. 26	High-Energy Physics and Nuclear Structure – 1975 (Santa Fe and Los Alamos)	75-26411	0-88318-125-8
No. 27	Topics in Statistical Mechanics and Biophysics: A Memorial to Julius L. Jackson (Wayne State University, 1975)	75-36309	0-88318-126-6
No. 28	Physics and Our World: A Symposium in Honor of Victor F. Weisskopf (M.I.T., 1974)	76-7207	0-88318-127-4
No. 29	Magnetism and Magnetic Materials – 1975 (21st Annual Conference, Philadelphia)	76-10931	0-88318-128-2
No. 30	Particle Searches and Discoveries – 1976 (Vanderbilt Conference)	76-19949	0-88318-129-0
No. 31	Structure and Excitations of Amorphous Solids (Williamsburg, VA, 1976)	76-22279	0-88318-130-4
No. 32	Materials Technology – 1976 (APS New York Meeting)	76-27967	0-88318-131-2
No. 33	Meson-Nuclear Physics – 1976 (Carnegie-Mellon Conference)	76-26811	0-88318-132-0
No. 34	Magnetism and Magnetic Materials – 1976 (Joint MMM-Intermag Conference, Pittsburgh)	76-47106	0-88318-133-9
No. 35	High Energy Physics with Polarized Beams and Targets (Argonne, 1976)	76-50181	0-88318-134-7
No. 36	Momentum Wave Functions – 1976 (Indiana University)	77-82145	0-88318-135-5
No. 37	Weak Interaction Physics – 1977 (Indiana University)	77-83344	0-88318-136-3
No. 38	Workshop on New Directions in Mossbauer Spectroscopy (Argonne, 1977)	77-90635	0-88318-137-1
No. 39	Physics Careers, Employment and Education (Penn State, 1977)	77-94053	0-88318-138-X
No. 40	Electrical Transport and Optical Properties of Inhomogeneous Media (Ohio State University, 1977)	78-54319	0-88318-139-8
No. 41	Nucleon-Nucleon Interactions – 1977 (Vancouver)	78-54249	0-88318-140-1
No. 42	Higher Energy Polarized Proton Beams (Ann Arbor, 1977)	78-55682	0-88318-141-X
No. 43	Particles and Fields – 1977 (APS/DPF, Argonne)	78-55683	0-88318-142-8
No. 44	Future Trends in Superconductive Electronics (Charlottesville, 1978)	77-9240	0-88318-143-6
No. 45	New Results in High Energy Physics – 1978 (Vanderbilt Conference)	78-67196	0-88318-144-4
No. 46	Topics in Nonlinear Dynamics (La Jolla Institute)	78-57870	0-88318-145-2
No. 47	Clustering Aspects of Nuclear Structure and Nuclear Reactions (Winnepeg, 1978)	78-64942	0-88318-146-0
No. 48	Current Trends in the Theory of Fields (Tallahassee, 1978)	78-72948	0-88318-147-9

No. 49	Cosmic Rays and Particle Physics – 1978 (Bartol Conference)	79-50489	0-88318-148-7
No. 50	Laser-Solid Interactions and Laser Processing – 1978 (Boston)	79-51564	0-88318-149-5
No. 51	High Energy Physics with Polarized Beams and Polarized Targets (Argonne, 1978)	79-64565	0-88318-150-9
No. 52	Long-Distance Neutrino Detection – 1978 (C.L. Cowan Memorial Symposium)	79-52078	0-88318-151-7
No. 53	Modulated Structures – 1979 (Kailua Kona, Hawaii)	79-53846	0-88318-152-5
No. 54	Meson-Nuclear Physics – 1979 (Houston)	79-53978	0-88318-153-3
No. 55	Quantum Chromodynamics (La Jolla, 1978)	79-54969	0-88318-154-1
No. 56	Particle Acceleration Mechanisms in Astrophysics (La Jolla, 1979)	79-55844	0-88318-155-X
No. 57	Nonlinear Dynamics and the Beam-Beam Interaction (Brookhaven, 1979)	79-57341	0-88318-156-8
No. 58	Inhomogeneous Superconductors – 1979 (Berkeley Springs, W.V.)	79-57620	0-88318-157-6
No. 59	Particles and Fields – 1979 (APS/DPF Montreal)	80-66631	0-88318-158-4
No. 60	History of the ZGS (Argonne, 1979)	80-67694	0-88318-159-2
No. 61	Aspects of the Kinetics and Dynamics of Surface Reactions (La Jolla Institute, 1979)	80-68004	0-88318-160-6
No. 62	High Energy e^+e^- Interactions (Vanderbilt, 1980)	80-53377	0-88318-161-4
No. 63	Supernovae Spectra (La Jolla, 1980)	80-70019	0-88318-162-2
No. 64	Laboratory EXAFS Facilities – 1980 (Univ. of Washington)	80-70579	0-88318-163-0
No. 65	Optics in Four Dimensions – 1980 (ICO, Ensenada)	80-70771	0-88318-164-9
No. 66	Physics in the Automotive Industry – 1980 (APS/AAPT Topical Conference)	80-70987	0-88318-165-7
No. 67	Experimental Meson Spectroscopy – 1980 (Sixth International Conference, Brookhaven)	80-71123	0-88318-166-5
No. 68	High Energy Physics – 1980 (XX International Conference, Madison)	81-65032	0-88318-167-3
No. 69	Polarization Phenomena in Nuclear Physics – 1980 (Fifth International Symposium, Santa Fe)	81-65107	0-88318-168-1
No. 70	Chemistry and Physics of Coal Utilization – 1980 (APS, Morgantown)	81-65106	0-88318-169-X
No. 71	Group Theory and its Applications in Physics – 1980 (Latin American School of Physics, Mexico City)	81-66132	0-88318-170-3
No. 72	Weak Interactions as a Probe of Unification (Virginia Polytechnic Institute – 1980)	81-67184	0-88318-171-1
No. 73	Tetrahedrally Bonded Amorphous Semiconductors (Carefree, Arizona, 1981)	81-67419	0-88318-172-X

No. 74	Perturbative Quantum Chromodynamics (Tallahassee, 1981)	81-70372	0-88318-173-8
No. 75	Low Energy X-Ray Diagnostics – 1981 (Monterey)	81-69841	0-88318-174-6
No. 76	Nonlinear Properties of Internal Waves (La Jolla Institute, 1981)	81-71062	0-88318-175-4
No. 77	Gamma Ray Transients and Related Astrophysical Phenomena (La Jolla Institute, 1981)	81-71543	0-88318-176-2
No. 78	Shock Waves in Condensed Matter – 1981 (Menlo Park)	82-70014	0-88318-177-0
No. 79	Pion Production and Absorption in Nuclei – 1981 (Indiana University Cyclotron Facility)	82-70678	0-88318-178-9
No. 80	Polarized Proton Ion Sources (Ann Arbor, 1981)	82-71025	0-88318-179-7
No. 81	Particles and Fields –1981: Testing the Standard Model (APS/DPF, Santa Cruz)	82-71156	0-88318-180-0
No. 82	Interpretation of Climate and Photochemical Models, Ozone and Temperature Measurements (La Jolla Institute, 1981)	82-71345	0-88318-181-9
No. 83	The Galactic Center (Cal. Inst. of Tech., 1982)	82-71635	0-88318-182-7
No. 84	Physics in the Steel Industry (APS/AISI, Lehigh University, 1981)	82-72033	0-88318-183-5
No. 85	Proton-Antiproton Collider Physics –1981 (Madison, Wisconsin)	82-72141	0-88318-184-3
No. 86	Momentum Wave Functions – 1982 (Adelaide, Australia)	82-72375	0-88318-185-1
No. 87	Physics of High Energy Particle Accelerators (Fermilab Summer School, 1981)	82-72421	0-88318-186-X
No. 88	Mathematical Methods in Hydrodynamics and Integrability in Dynamical Systems (La Jolla Institute, 1981)	82-72462	0-88318-187-8
No. 89	Neutron Scattering – 1981 (Argonne National Laboratory)	82-73094	0-88318-188-6
No. 90	Laser Techniques for Extreme Ultraviolt Spectroscopy (Boulder, 1982)	82-73205	0-88318-189-4
No. 91	Laser Acceleration of Particles (Los Alamos, 1982)	82-73361	0-88318-190-8
No. 92	The State of Particle Accelerators and High Energy Physics (Fermilab, 1981)	82-73861	0-88318-191-6
No. 93	Novel Results in Particle Physics (Vanderbilt, 1982)	82-73954	0-88318-192-4
No. 94	X-Ray and Atomic Inner-Shell Physics – 1982 (International Conference, U. of Oregon)	82-74075	0-88318-193-2
No. 95	High Energy Spin Physics – 1982 (Brookhaven National Laboratory)	83-70154	0-88318-194-0
No. 96	Science Underground (Los Alamos, 1982)	83-70377	0-88318-195-9

No. 97	The Interaction Between Medium Energy Nucleons in Nuclei – 1982 (Indiana University)	83-70649	0-88318-196-7
No. 98	Particles and Fields – 1982 (APS/DPF University of Maryland)	83-70807	0-88318-197-5
No. 99	Neutrino Mass and Gauge Structure of Weak Interactions (Telemark, 1982)	83-71072	0-88318-198-3
No. 100	Excimer Lasers – 1983 (OSA, Lake Tahoe, Nevada)	83-71437	0-88318-199-1
No. 101	Positron-Electron Pairs in Astrophysics (Goddard Space Flight Center, 1983)	83-71926	0-88318-200-9
No. 102	Intense Medium Energy Sources of Strangeness (UC-Sant Cruz, 1983)	83-72261	0-88318-201-7
No. 103	Quantum Fluids and Solids – 1983 (Sanibel Island, Florida)	83-72440	0-88318-202-5
No. 104	Physics, Technology and the Nuclear Arms Race (APS Baltimore –1983)	83-72533	0-88318-203-3
No. 105	Physics of High Energy Particle Accelerators (SLAC Summer School, 1982)	83-72986	0-88318-304-8
No. 106	Predictability of Fluid Motions (La Jolla Institute, 1983)	83-73641	0-88318-305-6
No. 107	Physics and Chemistry of Porous Media (Schlumberger-Doll Research, 1983)	83-73640	0-88318-306-4
No. 108	The Time Projection Chamber (TRIUMF, Vancouver, 1983)	83-83445	0-88318-307-2
No. 109	Random Walks and Their Applications in the Physical and Biological Sciences (NBS/La Jolla Institute, 1982)	84-70208	0-88318-308-0
No. 110	Hadron Substructure in Nuclear Physics (Indiana University, 1983)	84-70165	0-88318-309-9
No. 111	Production and Neutralization of Negative Ions and Beams (3rd Int'l Symposium, Brookhaven, 1983)	84-70379	0-88318-310-2
No. 112	Particles and Fields – 1983 (APS/DPF, Blacksburg, VA)	84-70378	0-88318-311-0
No. 113	Experimental Meson Spectroscopy – 1983 (Seventh International Conference, Brookhaven)	84-70910	0-88318-312-9
No. 114	Low Energy Tests of Conservation Laws in Particle Physics (Blacksburg, VA, 1983)	84-71157	0-88318-313-7
No. 115	High Energy Transients in Astrophysics (Santa Cruz, CA, 1983)	84-71205	0-88318-314-5
No. 116	Problems in Unification and Supergravity (La Jolla Institute, 1983)	84-71246	0-88318-315-3
No. 117	Polarized Proton Ion Sources (TRIUMF, Vancouver, 1983)	84-71235	0-88318-316-1

No. 118	Free Electron Generation of Extreme Ultraviolet Coherent Radiation (Brookhaven/OSA, 1983)	84-71539	0-88318-317-X
No. 119	Laser Techniques in the Extreme Ultraviolet (OSA, Boulder, Colorado, 1984)	84-72128	0-88318-318-8
No. 120	Optical Effects in Amorphous Semiconductors (Snowbird, Utah, 1984)	84-72419	0-88318-319-6
No. 121	High Energy e^+e^- Interactions (Vanderbilt, 1984)	84-72632	0-88318-320-X
No. 122	The Physics of VLSI (Xerox, Palo Alto, 1984)	84-72729	0-88318-321-8
No. 123	Intersections Between Particle and Nuclear Physics (Steamboat Springs, 1984)	84-72790	0-88318-322-6
No. 124	Neutron-Nucleus Collisions – A Probe of Nuclear Structure (Burr Oak State Park - 1984)	84-73216	0-88318-323-4
No. 125	Capture Gamma-Ray Spectroscopy and Related Topics – 1984 (Internat. Symposium, Knoxville)	84-73303	0-88318-324-2
No. 126	Solar Neutrinos and Neutrino Astronomy (Homestake, 1984)	84-63143	0-88318-325-0
No. 127	Physics of High Energy Particle Accelerators (BNL/SUNY Summer School, 1983)	85-70057	0-88318-326-9
No. 128	Nuclear Physics with Stored, Cooled Beams (McCormick's Creek State Park, Indiana, 1984)	85-71167	0-88318-327-7
No. 129	Radiofrequency Plasma Heating (Sixth Topical Conference, Callaway Gardens, GA, 1985)	85-48027	0-88318-328-5
No. 130	Laser Acceleration of Particles (Malibu, California, 1985)	85-48028	0-88318-329-3
No. 131	Workshop on Polarized ^{3}He Beams and Targets (Princeton, New Jersey, 1984)	85-48026	0-88318-330-7
No. 132	Hadron Spectroscopy–1985 (International Conference, Univ. of Maryland)	85-72537	0-88318-331-5
No. 133	Hadronic Probes and Nuclear Interactions (Arizona State University, 1985)	85-72638	0-88318-332-3
No. 134	The State of High Energy Physics (BNL/SUNY Summer School, 1983)	85-73170	0-88318-333-1
No. 135	Energy Sources: Conservation and Renewables (APS, Washington, DC, 1985)	85-73019	0-88318-334-X
No. 136	Atomic Theory Workshop on Relativistic and QED Effects in Heavy Atoms	85-73790	0-88318-335-8
No. 137	Polymer-Flow Interaction (La Jolla Institute, 1985)	85-73915	0-88318-336-6
No. 138	Frontiers in Electronic Materials and Processing (Houston, TX, 1985)	86-70108	0-88318-337-4
No. 139	High-Current, High-Brightness, and High-Duty Factor Ion Injectors (La Jolla Institute, 1985)	86-70245	0-88318-338-2

No. 140	Boron-Rich Solids (Albuquerque, NM, 1985)	86-70246	0-88318-339-0
No. 141	Gamma-Ray Bursts (Stanford, CA, 1984)	86-70761	0-88318-340-4
No. 142	Nuclear Structure at High Spin, Excitation, and Momentum Transfer (Indiana University, 1985)	86-70837	0-88318-341-2
No. 143	Mexican School of Particles and Fields (Oaxtepec, México, 1984)	86-81187	0-88318-342-0
No. 144	Magnetospheric Phenomena in Astrophysics (Los Alamos, 1984)	86-71149	0-88318-343-9
No. 145	Polarized Beams at SSC & Polarized Antiprotons (Ann Arbor, MI & Bodega Bay, CA, 1985)	86-71343	0-88318-344-7
No. 146	Advances in Laser Science–I (Dallas, TX, 1985)	86-71536	0-88318-345-5
No. 147	Short Wavelength Coherent Radiation: Generation and Applications (Monterey, CA, 1986)	86-71674	0-88318-346-3
No. 148	Space Colonization: Technology and The Liberal Arts (Geneva, NY, 1985)	86-71675	0-88318-347-1
No. 149	Physics and Chemistry of Protective Coatings (Universal City, CA, 1985)	86-72019	0-88318-348-X
No. 150	Intersections Between Particle and Nuclear Physics (Lake Louise, Canada, 1986)	86-72018	0-88318-349-8
No. 151	Neural Networks for Computing (Snowbird, UT, 1986)	86-72481	0-88318-351-X
No. 152	Heavy Ion Inertial Fusion (Washington, DC, 1986)	86-73185	0-88318-352-8
No. 153	Physics of Particle Accelerators (SLAC Summer School, 1985) (Fermilab Summer School, 1984)	87-70103	0-88318-353-6
No. 154	Physics and Chemistry of Porous Media—II (Ridge Field, CT, 1986)	83-73640	0-88318-354-4
No. 155	The Galactic Center: Proceedings of the Symposium Honoring C. H. Townes (Berkeley, CA, 1986)	86-73186	0-88318-355-2
No. 156	Advanced Accelerator Concepts (Madison, WI, 1986)	87-70635	0-88318-358-0
No. 157	Stability of Amorphous Silicon Alloy Materials and Devices (Palo Alto, CA, 1987)	87-70990	0-88318-359-9
No. 158	Production and Neutralization of Negative Ions and Beams (Brookhaven, NY, 1986)	87-71695	0-88318-358-7

No. 159	Applications of Radio-Frequency Power to Plasma: Seventh Topical Conference (Kissimmee, FL, 1987)	87-71812	0-88318-359-5
No. 160	Advances in Laser Science–II (Seattle, WA, 1986)	87-71962	0-88318-360-9
No. 161	Electron Scattering in Nuclear and Particle Science: In Commemoration of the 35th Anniversary of the Lyman-Hanson-Scott Experiment (Urbana, IL, 1986)	87-72403	0-88318-361-7
No. 162	Few-Body Systems and Multiparticle Dynamics (Crystal City, VA, 1987)	87-72594	0-88318-362-5
No. 163	Pion–Nucleus Physics: Future Directions and New Facilities at LAMPF (Los Alamos, NM, 1987)	87-72961	0-88318-363-3
No. 164	Nuclei Far from Stability: Fifth International Conference (Rosseau Lake, ON, 1987)	87-73214	0-88318-364-1
No. 165	Thin Film Processing and Characterization of High-Temperature Superconductors	87-73420	0-88318-365-X